A Short History
of Physics in the
American Century

NEW HISTORIES OF SCIENCE, TECHNOLOGY, AND MEDICINE
Series Editors
Margaret C. Jacob
Spencer R. Weart

A Short History
of Physics in the
American Century

DAVID C. CASSIDY

HARVARD UNIVERSITY PRESS
Cambridge, Massachusetts, and London, England
2011

Library of Congress Cataloging-in-Publication Data
Cassidy, David C., 1945–
   A short history of physics in the American century / David C. Cassidy.
      p.   cm.—(New histories of science, technology, and medicine)
   Includes bibliographical references and index.
   ISBN 978-0-674-04936-9 (alk. paper)
   1. Physics—United States—History.   2. Physicists—United States.   I. Title.
   QC9.U5C36   2011
   530.0973'0904—dc22   2011004099

Contents

A Short History
of Physics in the
American Century

Introduction

Physicists were in a celebratory mood at the turn to the twenty-first century. Nearly every year of the decade straddling 2000 brought celebrations of discoveries and breakthroughs that had revolutionized physics and transformed the world in which we live. The hundredth anniversaries of the discoveries of X-rays (1895), radioactivity (1896), the electron (1897), the idea of the quantum (1900), and Albert Einstein's "year of miracles" (1905) fell during the years from 1995 to 2005.

Those years also encompassed the seventy-fifth anniversary of the breakthrough to quantum mechanics (1925), the sixtieth anniversary of the start of the Manhattan Project (1942), the fiftieth anniversary of the transistor (1947), the fortieth anniversary of the laser (1960), and the tenth anniversary of high-temperature superconductivity (1986). The hundredth anniversaries of the founding of the premier research journal *Physical Review* (1893) and the American physicists' main professional organization, the American Physical Society (1899), also fell within the last years of the century, marking milestones in the story of American physics during the past century.

Many of these more recent developments and discoveries had occurred within the United States and within a discipline of research that itself had undergone a remarkable transformation. During the twentieth century, American physics followed an arching trajectory that extended from relatively humble beginnings in 1900 to a place on the world stage by the early 1930s, to a place of preeminence by the 1950s, and finally by the end of the century to a more modest place as an internationalized discipline within a global community of competitors. This is a book about that trajectory: the people, the discoveries, the institutions. It includes both the well-known and the less well-known physicists and institutions, as well as often neglected minority participants. Here "American physics" refers to physics in the United States of America, the profession and the science.

The trajectory of American physics did not proceed in isolation but closely tracked and sometimes led the international growth, expansion, and shifts occurring within physics itself before and during the century. It also closely followed the trajectory of American history. By 1900 the United States was poised for rapid growth and influence. The nation had recently achieved industrialization, and its economy was primed for expansion. At the same time, as new discoveries emerged in Europe, new technologies reached commercialization in the United States, helping spur the demand for home-grown discoveries. The nation's influence grew as it became a key player on the world stage during the first and second world wars, and as it became the main competitor with the Soviet Union during the decades of the Cold War, which ended in the early 1990s. As the United States emerged from the world wars as one of the most powerful economic, military, and scientific nations on the planet, some of its leaders believed that its time had come to proclaim "the American Century," a century that they believed would be dominated, in rather imperial terms, by the spread of American democracy, culture, industry, and influence across the globe, all supported by the world's finest innovators in science and technology.[1] Among those innovators were, of course, the physicists who had helped win the war with radar and the atomic bomb. As the Cold War concluded during the last decade of the century, the United States found itself the last remaining superpower. But, after relinquishing its postwar commitment to scientific leadership early in the last third of the century, the nation also found itself struggling to maintain the competitive quality of its education and to foster its increasingly internationalized research base within a global environment populated by other industrial and industrializing nations intent upon raising their scientific expertise to equal or greater status.

How did all this happen? What forces were at work in the United States and within physics and the physics profession that brought about these changes in the status of its science?

What we call science is the body of knowledge about nature that has been assembled, tested, and constantly retested and revised by many generations of people in many different nations and cultures. It is, in the end, a product of our universal curiosity and drive to comprehend the natural world in which we live. But it is also a product of its time and of the people and institutions and cultures that produce it. By its name, physics focuses on the physical aspects of nature, the search for an understanding of the laws of nature that govern the properties and behavior of physical matter. As the properties, behaviors, and types of matter under investigation expanded and shifted over time and for a variety of reasons, the physics discipline itself expanded dramatically over the course of the century. This led to the growth of new, specialized subdisciplines, such as condensed-matter, plasma, medical, chemical, and computational physics, and to new "hyphenated" fields such as astrophysics, geophysics, and biophysics.

Because of the human quality of science and of the need for constant cooperation, testing, and revising, a nation's success in science arises from its ability to create a community of trained scientists and to support and encourage that community by creating the conditions that will foster research on the forefront of knowledge. Among these conditions are a steady supply of motivated young people prepared for the rigors of training and future research; ready access to higher education; sufficient funding, equipment, and professional standards; freedom to publish and debate results; and, for a newly emerging community, a widely accepted vision of its work and goals that enables it to create and maintain an independent professional identity within science and the broader society.

By itself, science produces only knowledge about nature. It does not directly produce any new goods or commercial products. But modern scientific research is expensive, highly technical, and often frustrating in that useful results are never guaranteed. As have scientists throughout history, American physicists found it necessary to rely for research support, social recognition, and training and employment on partnerships with other, more powerful institutions within society. Among them were government, industry, military, philanthropy, education, and the general public. And these institutions responded as they recognized—and were encouraged to recognize—at least two advantages that a healthy science, and physics, can bring. The first was best captured in Francis Bacon's famous statement that "knowledge is power": the power to produce new technologies leading to new consumer products, new instruments of research, new medical applications, new military weapons, and new solutions to national problems such as energy resources. A second advantage of a thriving physics is that it can be regarded as one component of a nation's culture and learning and thus as a valued indicator of a nation's cultural standing. This was especially important during the Cold War, a competition that both sides regarded as a clash, not just between two superpowers, but between two political, social, and economic systems whose strength and superiority were exemplified by the strength and superiority of their science and culture.

The history of American physics during the twentieth century is, of course, a very big subject and one that can be approached from many different perspectives. In addition, a great deal has already been written about the history of American physics during the past century—from histories of industrial facilities, philanthropic foundations, and accelerator laboratories to detailed studies of important discoveries, developments, and people; to accounts of defining events such as the Manhattan Project and radar development; to sociological, anthropological, and philosophical analyses of physicists and their work; to end-of-century surveys of advances in physics during the past century; to scholarly histories of the overall growth of the American physics discipline and science. Many of these studies, some published decades ago, are still considered authoritative within the spaces they occupy. They include, in particular, works by Daniel Kevles, A. Hunter Dupree, Richard Rhodes, and many of the other authors listed in the notes to this book.

Although it was first published in 1971, Kevles's *The Physicists: The History of a Scientific Community in Modern America* remains the standard historical account of the American academic physics profession from the late nineteenth century through the first two-thirds of the twentieth century. Dupree's *Science in the Federal Government*, published in 1957, is a history of federally sponsored research, including physics, from the Jeffersonian era to the eve of the nation's entry into World War II.[2] Neither author fully explores research in industrial laboratories. Dupree ends his account just before the influx of massive federal funding for physics beginning with the atomic bomb and radar development and continuing into the Cold War decades. While Kevles's account covers the postwar federal sponsorship of physics, including military funding, through the peak decades of the 1950s and 1960s, he tends to underestimate the ways in which this massive funding helped shape the direction, and possibly even the content, of physics research. His account ends just before the rapid decline of federal funding during the last third of the century, paralleled by the rise of industrial funding to its now prominent position. This book attempts to encompass all three of these main research venues over the course of the entire century.

Nevertheless, this book is intended only as a very brief introductory synthesis of the history of twentieth-century American physics for students and the general public. It is not intended to offer a new analysis of that history or to argue a newly constructed thesis. In keeping with its goals and audience, this book does not drift far from the standard, often currently definitive literature on its subject—as far as that literature goes. My own supplementary research, reflection, and interpretation are included throughout, especially regarding the last third of the century. Because of the brevity of this book, many topics could not be included, and those that could be are condensed. As an aid to further study, readers are referred to the works listed in the corresponding notes for fuller accounts, if available, of events, developments, and technical matters discussed in the text.

Despite the many end-of-century celebrations, I have sought to portray the profession and its developments and discoveries not merely as a progression of important figures and events marching inexorably toward success, but rather as together forming one important component of the broader fabric of national and international historical development that, with its many struggles, failures, and successes, helped comprise what we call twentieth-century history. However much one may admire the achievements of physics over the past century, the task of history is not to celebrate achievements but to understand them as historical events, to explore and explain what happened, how it happened, and why it happened.

1
Entering the New Century

The discoveries emerging from the European powerhouses of physics during the 1890s heralded the approach of the new century and helped set the stage for the future century of physics. Taking advantage of recent advances in electromagnetic theory and precision instrumentation, Cambridge University physicist J. J. Thomson discovered the first subatomic particle, the electron. In Paris, Henri Becquerel discovered radioactivity, and soon Marie and Pierre Curie uncovered new radioactive elements that would win them and Becquerel Nobel Prizes. In the Netherlands, H. A. Lorentz developed a new theory of electromagnetism, and Pieter Zeeman discovered the Zeeman Effect, the magnetic separation of atomic spectroscopic lines.

In 1900 Max Planck, working in Berlin, hypothesized the existence of tiny, indivisible quantities, or quanta, of energy. Five years later, Albert Einstein, an unknown clerk in the patent office in Bern, Switzerland, introduced the theory of relativity, the quantum theory of light, the equivalence of mass and energy, and the theoretical basis for confirming the existence of atoms. In 1907 he presented a quantum theory of crystal solids, and he began work on a general theory of relativity designed to encompass gravitation. In 1909, in a brilliant series of experiments, Ernest Rutherford and his assistants at the University of Manchester in England first discovered the atomic nucleus. And in 1913 Danish physicist Niels Bohr published the first quantum theory of atomic structure.

American contributions could not compare with those achievements. While Europeans produced the breakthroughs that helped launch the new twentieth-century disciplines of relativity theory and atomic and quantum physics, American researchers were at work during the early years of the century on such standard, "classical" topics as electromagnetism, optics, acoustics, and electrical and thermal properties of materials. During the first decade of the century, they produced such highlights as the first observation of the radiation pressure of light, the magnetic rotation of sodium vapor, and measurements of the heat developed in a material due to radioactivity. Of the fifty-four articles appearing in the ten issues of *Physical Review* published during 1900, the most popular category, accounting for nineteen of the articles, concerned the development of experimental apparatus for electrical research, teaching, and industry. The next most popular general topic, with sixteen articles, entailed the study of electromagnetic, thermal, and other properties of matter. Electricity and magnetism appeared as independent topics in only five papers. Thermodynamics and the physics of air and gases accounted for three articles each, while acoustics, optics, radioactivity, spectroscopy, the photoelectric effect, and units accounted for the rest.[1]

Despite the uneven comparison with European research, the United States did produce several big-name physicists during the late nineteenth century whose work did rival that of their European colleagues. J. Willard Gibbs at Yale University developed independently a statistical mechanics of heat and gases that was comparable to the great Austrian and German theories put forth at that time by Ludwig Boltzmann and Rudolf Clausius. Albert Michelson, together with Edward Morley at what is now Case Western Reserve University, undertook precision optical interference experiments, which in 1887 yielded strong evidence against the electromagnetic ether, the hypothetical medium for light waves. Michelson received the first American Nobel Prize for this and related work in 1907. Henry Rowland, the first professor of physics at Johns Hopkins University, was known internationally for his precision measurements of physical constants and for his invention of the widely used Rowland diffraction grating for optical research.

But these outstanding figures were the exceptions, and their work did not spark many of their American colleagues to similar action at that time. Rowland died in 1901 and Gibbs in 1903. Neither was replaced by a successor of equal stature. Michelson, however, became professor of physics at the University of Chicago in 1892 where he helped establish a first-rate research program.

Four events at or near the turn of the century signaled the transformation in American physics that was about to occur. In 1893, three physicists at Cornell University founded the *Physical Review*, a journal that in the decades ahead would become the world's leading physics research journal. In 1899, thirty-six physicists meeting at Columbia University founded the American Physical Society. In 1900 the General Electric Company established the nation's first major industrial research laboratory. And in 1901 the federal government created the National Bureau of Standards under the direction of a physicist. Its purpose was to set standards for industrial products and government regulations on the basis of fundamental research. These events indicated not only that American physicists were beginning to assume an independent professional identity within American science and society, but also that they were beginning to obtain through industry and government as well as academe the direct support for their work that they would need to bring their profession to world stature during the years ahead. It was still only a beginning, but these events were the culmination of a series of important developments reaching back to earlier decades and even to earlier centuries.

TWO TECHNOLOGIES

Science and physics especially had always been prized fixtures of American culture. Benjamin Franklin's discoveries and inventions in electricity and other fields were only the most prominent of many others. Inspired by the Enlightenment ideals of government guided by reason, natural law, and the practical benefits of science, all of the Founding Fathers held science in high esteem, even if science was mainly accessible only to a few wealthy, white male aristocrats.

The Civil War proved a turning point in American social and science history. Following the devastation of the war, the population rapidly expanded as the surviving combatants returned to civilian life and as European immigrants began pouring into the United States to seek a better future. The nation badly needed workers for its rapidly expanding economy. The nation also needed more people trained in advanced agricultural methods to support the growing population, and in engineering to support the industrial revolution as it gained momentum during those decades. Federal funds began flowing through the Morrill Act of 1862 primarily to western and midwestern states for the establishment of practical training in the "agricultural and mechanic arts." The new funds stimulated the founding of numerous public land-grant colleges, many of which are now prominent state universities, and those colleges began producing a new educated middle class.

Two technological developments spurred the rise of the nation's industrial economy and with it the nation's science and engineering. The first was the harnessing of the steam engine to manufacturing and the railroad. The completion of the transcontinental railroad in 1869 opened up the American continent to commercial trade. This vastly increased the need for more trained mechanical engineers to design and build improved steam engines.

The second development entailed the onset of the electric age following the invention of the electric motor in Europe in 1873. (The development of the generator, or dynamo, had occurred in 1830.) During the next twenty years, numerous inventions literally electrified the nation, driving the public demand for more electric power. Steam turbines, consuming vast quantities of fossil fuels, were harnessed to generate electricity; the Edison Lighting Company began selling Thomas Edison's incandescent light bulb and the electric networks to run them; and the country finally settled on an electric standard of 110-volt alternating current (AC) running at 60 cycles per second. As with the steam engine, the electric age required trained experts, this time in the new and highly technical profession of the electrical engineer.

The rise of chemistry during the nineteenth century and of the chemical and petroleum industries, along with other science-based industries, also required greater technical expertise and training in physical science.

At the same time, midwestern states, eager to achieve industrial development, began requiring their land-grant colleges to offer, in additional to agricultural sciences, courses for future engineers. Many of those colleges gained closer ties with the railroad and electrical industries that employed the newly trained engineers. And at the base of every engineering student's education lay a firm foundation in the fundamentals of physics—electricity, magnetism, thermodynamics, and mechanics.[2]

The increase in the number of students and courses drove an increasing demand for undergraduate physics professors. But physics was not yet an independent discipline. Usually, as at the land-grant college Purdue University in Indiana, which may be regarded as representative, the physics professors were located in the electrical engineering department. There they taught basic physics to the future designers of industrial-age machines. According to the Purdue course catalogue for 1895–1896, physics was divided into two subjects: general and practical physics. The annual courses in practical physics were "devoted to precise physical measurements" of various quantities using instruments common to engineering practice. The lectures in general physics covered mechanics, heat, and sound, with heavy emphasis in the senior year on electricity and magnetism. But, in a sign of awakening independence, attention was also given "to the recent advances in physical science," and in their senior year students were even given "a subject for original investigation" that could serve as their senior thesis.[3]

However, most physics professors, burdened with heavy teaching loads, did not themselves perform any original research. During the 1890s, there were about 200 practicing physicists in the United States. Most of them held only a bachelor's degree, even if teaching in a college. Unlike today, only about one-fourth of practicing physicists held a doctoral degree, usually from a European university. Purdue University's professor of physics from 1899 to 1928 held a bachelor's degree from Cornell University, followed by a year of graduate study at Uppsala University in Sweden. In contrast, nearly all European physics professors held doctorates (although the requirements were somewhat less demanding); most were working in dedicated university research laboratories staffed by graduate assistants; and they were in the classroom less and in the laboratory more than their American counterparts.[4]

The special success of the European graduate research system served for many years as an ideal for American scientists, and often as a source of envy. In 1876, American educators, eager to import the European methods, founded Johns Hopkins University following the German model of the graduate research university. Graduate research programs in physics at Cornell, Harvard, Yale, and a few other universities eventually followed. They produced nearly all of the American-conferred doctorates in physics, and they helped to account for the fact that about 20 percent of American physicists did manage to find some time to perform research.[5] Most of the research involved the empirical observations and problems of practical concern noted earlier in their publications of 1900, rather than involving the search for fundamental new concepts and discoveries. The discipline still lacked definition and the standards and mechanisms needed for ensuring quality research.

THE ROLE OF FEDERAL LABORATORIES

At the turn of the century, the heaviest concentration of research physicists was found not in universities but in the federally funded national research laboratories established in Washington, D.C.[6] Federal research establishments also went back to the Founding Fathers, and to the immediately following century. Established in 1844 by an act of Congress with the initial funding of British scientist James Smithson, the Smithsonian Institution in Washington, D.C., served for decades as the premier sponsor of research in botany, anthropology, and geographic exploration. When solar physicist Samuel Pierpont Langley became secretary of the Smithsonian in 1901, he built a renowned astrophysical observatory— one of the first instances of the joining of physics with another discipline, astronomy, to form a new hybrid field of study: astrophysics. After the Wright brothers' success, Langley became a leading proponent of aviation research.

By the dawn of the twentieth century, the nation had become industrialized and urbanized. But there were as yet no industrial or academic research laboratories able to establish guidelines and standards for technology-based industries or for the management of a rapidly expanding population. These matters were left to the federal government which, in addition to the National Bureau of Standards, established the Census Bureau, the Bureau of Mines, and the Office of Weights and Measures—all headed by research scientists. The Office of Weights and Measures, modeled after a German institute, worked to enable American industries to compete directly with their European counterparts in the sale of precision instruments and technology-based consumer products.[7]

One of the effects of the founding of government laboratories was the closer alliance of academic physicists with government and industry. The context for this, and one of the driving forces behind the government support of research, was the confluence of two social causes at that time: the progressive and conservation movements. Both of these coalesced during the presidency of Theodore Roosevelt (1901–1909). But even before then, as the nation industrialized, newly middle-class families sought to maintain their status by placing their children in professional careers through higher education. College enrollments nearly doubled during the 1890s, and they continued to grow in the decades ahead, although the numbers still remained relatively small. The 82,000 registered college students in 1900 represented only about 2.4 percent of the total college-age population.[8]

Many of the colleges followed and promoted the Progressive Movement's vision of science education inspired by the ideals of the Enlightenment. According to this vision, a new generation of technical experts trained in the latest science and engineering would readily solve the nation's problems by applying rational, scientific methods and results to their solution. This put scientists on a mission to serve the public and the nation by finding the best applications of science to such tasks as making more rational and efficient use of natural resources and the development of new technology-based consumer products and standards for industrial mass production. This in turn brought scientists into a closer collaboration with partners in both government and industry.

Government officials believed that an enlightened, "progressive" government should utilize its scientists to the benefit of the nation, and that it should support them in their work through the establishment of national laboratories. For the conservation of natural resources, programs initiated by the Bureau of Land Management, the Coastal Survey, and the Smithsonian were already under way and more initiatives occurred through 1916, all looking to science for support and guidance. Before industries began founding their own laboratories, the federal government, through its various research units, took over the task of solving complex technology and standards problems for them, often with funding from industry. The government provided most of the research for the nascent aviation industry before World War I, and it researched and set many of the technical standards for the American electrical industry. Because of this, the federal government took a decisive role in the course of industrial and commercial development during that era and in fostering physics and other sciences in support of that goal.[9]

UNIVERSITIES AND PURE SCIENCE

The fragile alliance of government, industry, and academe in scientific research grew more fragile as some corporations became uneasy about the government's influence on applied research.[10] As an alternative, beginning in the 1880s through the turn of the century, what became the Massachusetts Institute of Technology (MIT) received funds from its state, private sponsors, and industries to found a series of academic laboratories for electrical engineering, electrochemistry, and other industry-related fields. Unlike applied research, which seeks to turn physical concepts and laws into practical uses, engineering combines science, practical need, and industrial processes and design into a technology that can lead to useful products. The MIT laboratories became the basis for an undergraduate engineering program that trained its students in close collaboration with the needs of the institute's industrial sponsors. Similar partnerships sprang up elsewhere, but they also raised questions about the role of industry in sponsoring academic research, including research in physics. In particular, in what ways might the aims of outside sources of funding exert an influence on the internal course and even the content of research?

Other questions arose regarding research in government settings. In 1903 a federal committee appointed by the president, called the Committee of Organization of Government Scientific Work, struggled with the mission of government research. Should federal laboratories also engage in "problem oriented" research at the frontiers of knowledge, or should they limit themselves to practical applications of existing knowledge for the benefit of industry? It was a question that has occupied many agencies and scientists ever since. And it raises a larger question: what exactly is the relationship between fundamental, basic, or "pure" research and applied, practical, or useful research? More succinctly, what is the relationship between science and technology? As we will see, these questions were answered in different ways throughout the century. The 1903 committee decided that, because federal research is funded by Congress, it should focus on the solution of practical problems of direct benefit to the nation and to industry, whereas "disinterested pure science" should be left to the nonprofit universities.[11]

But universities already found themselves in a dilemma. On one side, industry was providing growing support of applied research in university laboratories, and it was demanding better preparation of students for engineering careers. On the other side, most universities were governed by administrators who adhered to ideals derived from the Progressive Movement of that era, especially the notion that research should be regarded as a public service of benefit to everyone, independent of commercial interests and profits. It served the public by contributing to the nation's culture and by providing potential practical benefits that could not yet be foreseen. Moreover, enhancing basic research would help bring the nation into closer competition with European nations for both prestige and competitive commercial applications.

The dilemma opened a division regarding the aims and methods of physics education in the United States between the land-grant engineering-oriented colleges along with MIT on the one hand, and the universities offering graduate research degrees on the other. The division reflected a split over the aims and methods of the physics profession itself. Most physicists in technical colleges saw their work as a service to the engineering professions through the teaching of elementary and applied physics to future engineers. In contrast with this, many of those in graduate universities saw their work as a service to the nation and to humanity as well as to physics through the training of new generations of professional physicists able to advance our knowledge of nature through original "pure" research, even if this research was not yet up to European standards. Training students in fundamental research required a professor adept in such research, which in turn required a rationale for the financial support of work that might have no foreseeable practical benefit. One basis for this rationale was a definition of the work of the physicist as the pursuit of "pure" knowledge untainted by, and even superior to, practical and applied research. New pure knowledge was its own reward, it was argued, while the possibility always existed that others might eventually put that knowledge to practical use.

This line of reasoning, motivated by the practical goal of achieving professional identity, became known as the ideology of "pure science," the idea of research activity driven solely by curiosity and the hunt for the fundamental laws of nature, independent of the needs or demands of commercial, social, and political interests.[12] It strongly influenced physicists' attitudes toward academic science policy and toward their own research, as we shall see, whenever the profession needed strong self-definition, as in this period, or self-promotion in seeking partnerships with other sectors of society. In other words, it was as much an instrument for professional promotion as it was a standard of conduct for its practitioners. Its effects can be felt even today. But, one may ask, could "disinterested pure science" really remain free of commercial and political interests, especially as those interests provided the funds necessary for research? One might ask further about the responsibilities that pure scientists, in their independence, should have to society and to the uses that society makes of their work.[13] These were important questions that occupied scientists, policymakers, and the general public during the decades ahead, even as the idea of pure science served its purpose in this period by helping to define and promote the nation's physics discipline.

The notion of pure science was articulated most clearly and forcefully in the United States by Henry Rowland. The American-born Rowland had studied physics in Germany with the famous Hermann von Helmholtz before becoming professor of physics at Johns Hopkins University. His experience in Germany had impressed upon him the lowly status of physics in his homeland and the debilitating concern of American physicists with practical applications rather than with advancing the forefront of knowledge. Returning to the United States, Rowland delivered an address to the American Association for the Advancement of Science (AAAS) in 1883 titled "A Plea for Pure Science." "We are tired of mediocrity, the curse of our country," he declared. "We are tired of seeing our professors degrading their chairs by the pursuit of applied science instead of pure science; or sitting inactive while the whole world is open to investigation; lingering by the wayside while the problem of the universe remains unsolved . . . Nature calls to us to study her, and our better feelings urge us in the same direction."[14]

The roots of pure science lay in the Progressive Movement, but they extended back further to the eighteenth-century Enlightenment and even to ancient Greek philosophers, especially Plato and his Academy of elite thinkers. But the most direct origins arose from nineteenth-century German professors. Rowland's European sojourn was not uncommon for American physical scientists of the era. While immersing himself in German physics, he apparently also absorbed the professional tactics of his German colleagues. In promoting their social standing and professional support within the newly unified monarchy under Bismarck and the Kaiser, German professors had developed the highly successful argument that "apolitical" pure research was essential to advancing German national culture and prestige. Because of this, all academic work, including science, merited generous state funding, and all research professors deserved, and received, special social and economic status, placing them just below the landed aristocracy among the leading professions in Germany. To this day, German professors and teachers are accorded higher social status than is accorded their American counterparts.[15]

In a similar fashion, at a time when the poorly educated inventor Thomas A. Edison was considered the nation's leading scientist, Rowland attempted to separate "pure science" and its academic practitioners from consumer products and their inventors. "American science is a thing of the future, and not of the present or past," Rowland told the AAAS in his 1883 speech, "and the proper course of one in my position is to consider what must be done to create a science of physics in this country, rather than to call telegraphs, electric lights, and such conveniences, by the name of science." In order to answer nature's call to research, he continued, professors must place pure science above commercial pursuits, and they must find a way to engage in pure research, in addition to teaching, as do European professors. Moreover, universities must hire more physics professors who "choose a life which we consider higher" so that students, "looking forward into the world for something to do, see before them this high and noble life, and they see that there is something more honorable than the accumulation of wealth."[16] Like his German counterparts, Rowland used the idea of pure science as an ideology, as an instrument for the establishment and recognition of academic research physics as a recognized profession set apart from and above politics as well as above applied science and commercial technology.

In the same speech, Rowland also set forth a program of practical steps for fostering the new elite profession and its institutions. The steps included, first, the tapping only of private donors for funds in order to avoid, as did his German colleagues, the taint of "political trickery" associated with government support; second, the channeling of these funds primarily to a few selected elite universities with first-class graduate research programs in physics, leaving the majority to languish in mediocrity. In addition, he called for the founding of research journals specifically for physics research and the establishment of a professional society of research physicists to promote themselves and pure science and to set standards of excellence—goals realized with the founding the *Physical Review* and the American Physical Society (APS).

In 1899, Rowland was duly elected the first president of the APS, an organization that was populated primarily by academic physicists until after World War II. In 1913 the APS took editorial control of the *Physical Review*. Despite competition from other journals publishing physics research, such as *Science* (AAAS), the *Astrophysical Journal*, and the *Journal of Physical Chemistry*, the *Physical Review* served for the next several decades as a vehicle for the physicists' growing professional identity. Unlike today, during its first decades the *Physical Review* published not only research papers, but also reports and abstracts from APS meetings, book reviews, summaries of papers published elsewhere, obituaries, and advertisements. Yet, as Paul Hartman notes in his "memoir" of the journal, it was through the research published in its ever expanding volumes that the journal reflected and chronicled the growth of the American physics community over the next 100 years.[17]

APS president Rowland could not have been more pleased than to report to the society's second meeting in 1899 the success of his elitist vision of pure science: "In a country where the doctrine of the equal rights of man has been distorted to mean the equality of men in other respects, we form a small and unique body of men, a new variety of the human race . . . whose views of what constitutes the greatest achievement in life are very different from those around us." His German colleagues would have been proud as he continued: "In this respect we form an aristocracy, not of wealth, not of pedigree, but of intellect and of ideals, holding him in the highest respect who adds the most to our knowledge or who strives after it as the highest good."[18]

INDUSTRIAL LABORATORIES

As Rowland pushed academic physics toward independent professional status, industry was moving toward the creation of its own more commercially oriented research laboratories.[19] Several factors, again centering on the Progressive Movement, contributed to this trend. Progressivism encouraged young scientists entering the workforce to view industrial research as a contribution to advancing American economic and cultural stature. It did not hurt that scientists could earn much more in industry than in academe. Older scientists, however, imbued with notions of pure science, regarded this move as a sellout. Knowledge was for the benefit of all, they argued, not for the profits of a few. Frank Jewett, a physicist who received his doctorate from the University of Chicago in 1902, recalled that his dissertation advisor, Albert Michelson, "thought that when I entered industrial life, which was a field where patents were a part, I was prostituting my training and my ideals."[20]

New discoveries and processes possibly leading to new industrial applications constituted a second factor encouraging industrial research. Among these were breakthroughs in electrochemistry, electron physics, the conduction of electricity through gases, and surface phenomena, in addition to the properties of electromagnetic waves. These developments were potentially significant for two of the most actively developing technological industries at that time, the electric and telephone companies.

A third factor also arose from the progressive outlook: the antitrust movement of that era. Progressives believed that by breaking up the large corporations and monopolies that had emerged during the industrial revolution the government would encourage competition and innovation and ensure a more rational and efficient organization of the economy. But in that technological age, much like the computer and digital ages of today, discoveries potentially leading to new technologies could be essential to the survival of a company. So, too, was the control of the pace of new discoveries and technologies and of the spread of innovations to one's competitors, also similar to today. One way to control the pace of innovation and to keep new discoveries within the corporation, even if it was broken up into smaller companies, was to establish a semiautonomous central research laboratory at the forefront of pure research and to patent its discoveries as soon as they were made.[21]

Although the Standard Oil Company had opened a petroleum refining laboratory in 1882, in 1900 the General Electric Company (GE) established the nation's first industrial laboratory engaging in research not directly related to the production process. Under the direction of Willis Whitney, a chemist who had received his doctorate under the famed Wilhelm Ostwald at the University of Leipzig, the GE Research Laboratory encompassed a wide range of electromechanical technologies, including locomotives, dynamos, and motors. But its main focus was the investigation of physical and chemical processes occurring in the electric lamps developed by one of its founders, Thomas A. Edison. In 1909 Irving Langmuir, having received his doctorate in physical chemistry under Walther Nernst in Göttingen, joined the GE laboratory where he studied electric discharges from filaments in high-vacuum incandescent lamps. He later received the Nobel Prize in chemistry for his discoveries regarding surface chemistry.

In 1902 the DuPont Company founded its first laboratory. Many other industrial laboratories soon followed. But Rowland would have been disheartened to learn that, although the membership of APS had risen by 1913 to 682 members (from 36 in 1899), roughly 10 percent of those members were industrial physicists. Moreover, many of the research papers now appearing in *Physical Review* originated from industrial laboratories.[22]

Not to be outdone by GE, the American Telephone and Telegraph Company (AT&T), part of the Bell System, established its own research laboratory in 1911 in order, writes historian Leonard Reich, to preserve its market dominance in telecommunications. "The new research laboratory worked to advance and control those technologies."[23] In 1925 AT&T greatly expanded the laboratory and renamed it Bell Telephone Laboratories.

Bell Labs quickly became one of the nation's premier research laboratories, often performing pure research seemingly unrelated to the business of its sponsor. Many fundamental discoveries and theories, from transistors to cosmology, along with six Nobel Prize winners, emerged from Bell Labs during the course of the century. From the very start, the president of Bell Labs, Frank B. Jewett, had been one of the most creative and influential pioneers of AT&T research.

Jewett is remembered as a stately gentleman with an intellect that was "cool, clear, and penetrating, never impassioned or combative." Owing to a childhood illness, he was blind in one eye and seriously impaired in the other.[24] Descended from a long line of New England ancestors, Jewett was born in 1879 in Pasadena, California, where his family had recently settled, having purchased a 25-acre ranch in the vicinity. While managing the ranch, Jewett's father, a mechanical engineer trained at MIT, helped found the railroad line from Los Angeles to Pasadena. Young Frank attended a one-room school in Pasadena and grew understandably fascinated with railroads. He enrolled in the Throop Institute of Technology in the same town, which later became the California Institute of Technology (Caltech). After receiving his doctorate in physics with Michelson at the University of Chicago, Jewett taught for two years as a physics instructor at MIT. In 1904 he joined the Mechanical Department within the engineering section of AT&T in Boston, much to Michelson's displeasure, and in 1906 Jewett became director of the transmission research group within the department. A year later, the Mechanical Department merged with the Engineering Department of Western Electric, another member of the Bell System, and Jewett moved to New York City, where in 1915 he led the research and development effort that resulted in the first transcontinental telephone line.

By 1910, AT&T was eager to dominate wireless radio technology as well as wired long-distance telephone transmission. In order to do so and to achieve the crucial "repeater," a device required to boost the signal for transcontinental telephone calls, Jewett argued that a new research laboratory was needed "to employ skilled physicists who are familiar with the recent advances in molecular physics and who are capable of appreciating such further advances as are continually being made."[25] In 1910 AT&T's board of directors approved the establishment the following year of the

Research Branch within Western Electric's Engineering Department, with Jewett and a colleague, John J. Carty, as co-directors. This became the core unit of Bell Labs in 1925. The Engineering Department stated in its annual report in 1911, "There is an increasing number of problems intimately associated with the development of the telephone business . . . In the past, development work has proceeded more or less blindly by cut and try [sic] experimental methods. While these methods have given reasonable good results, it is felt that the time is ripe for investigation covering the fundamental principles."[26]

As co-director of the Research Branch, Jewett tapped his connections at Chicago, MIT, and other research universities to recruit a host of the best and brightest young physics doctorates for work in his new organization. Among them were Harold D. Arnold, Oliver E. Buckley, and H. W. Nichols. Not all succeeded in industrial research. Despite the higher pay, some regretted leaving the "pure science" of academe untainted by the drive for commercial profits. Apparently replying to this complaint, Jewett proclaimed, "The performance of industrial laboratories must be moneymaking . . . For this reason they cannot assemble a staff of investigators to each of whom is given a perfectly free hand."[27] However, as the United States entered the world war in 1917, Jewett argued in a speech that applied industrial research and pure science at universities must now grow together into a unification of science and technology. Each alone could not succeed. For the rest of his life, Jewett played a key role at Bell Labs and in government service in helping to realize this goal. He died in Summit, New Jersey, in 1949.

THE BUILDING BOOM

The founding of independent departments of physics at most colleges and universities at the turn of the century paralleled the appearance of independent industrial laboratories. Together they reflected the physicists' growing new professional identity. After a recession during the early 1890s, an economic boom filled the financial coffers of most states and the bank accounts of wealthy individuals, while a sudden increase in the number of students flooded the existing facilities. Beginning with the University of Chicago in 1894, the economic and population booms led to a construction boom of new university physics buildings funded by states and wealthy individuals through 1913 to house the new physics departments. Some of the new buildings were at private research universities, including Columbia (1897), Princeton (1909), and Yale (1912). Many others were at state land-grant colleges, including Minnesota (1902), Ohio State (1903), Purdue (1904), Michigan (1905, an extension), Illinois (1909), Iowa (1912), and Cornell (1905). (Cornell is both a private and a land-grant college, as well as a member of the Ivy League.)

State support of public education brought physics to many of the students at state colleges, with or without "political trickery." As private foundations and individuals funded buildings at the few graduate physics universities, Rowland's hierarchy of undergraduate practical physics at colleges versus elite pure graduate research at select universities was emerging by default.[28] Future engineers educated at a land-grant college did not require advanced graduate physics for their careers. Those who wanted to go into pure-physics research could head to Chicago or to one of the eastern graduate schools for advanced training. Those who did, such as Frank Jewett, were often able to do so only because their wealthy families could afford it. The aristocratic character of elite physics was still in place. Not until the 1920s did most land-grant colleges begin adding graduate physics programs to their curricula.

Reflecting the two-tiered approach, physics in the newly constructed department buildings at state colleges still focused mainly on undergraduate courses for future engineers. But they now also included preparation of those intending to meet the rising demand for high-school physics teachers. For instance, Purdue University established an independent physics department with its own building in 1904. The building possessed advanced teaching laboratories for senior projects and, for the first time, laboratories for the special use of future teachers. Also a first, the department began offering a two-semester course titled "theoretical physics." But the most popular courses were still those in "practical physics," for which there were three dedicated laboratories: one each for heat studies for mechanical engineering, electrical measurements for electrical engineering, and acoustics for telephone technology.[29]

Meanwhile, on the upper tier, the graduate programs at just five elite schools—Johns Hopkins, Cornell, Harvard, Yale, and Chicago—were producing the majority of work in pure research and the majority of physics doctorates. The numbers were increasing rapidly. In 1900 the country possessed a total of only fifty-four PhDs in physics; by 1909, it was producing twenty-five new PhDs every year. But still, by one count, only one-eighth of American physicists were publishing more than one-half of the nation's research in physics.[30]

Even below the top tier of researchers, most American physicists were aware of the burst of discoveries in experimental research occurring in Europe beginning in the 1890s. But few took much notice of the new theoretical developments surrounding the relativity and quantum theories. And few Europeans took note of American developments. Although in 1900 the United States had more physicists than did Germany, the quality of German research was much higher, and Germany did lead the world in one crucial area—it possessed sixteen theoretical physicists in 1900 compared with only three in the United States.[31] While most American physicists were still focusing on elementary and practical physics, European physicists were producing the breakthroughs that would drive physics research during the century ahead.

By the outbreak of World War I in 1914, European physicists were still far ahead of Americans in both experimental and theoretical work. But that was already slowly changing. In 1899, Albert Michelson had brought American physicist Robert A. Millikan back from Germany to a professorship at the University of Chicago. Michelson's Nobel Prize for physics in 1907 was followed in 1909 at the University of Chicago by Millikan's measurement of the charge on the electron. In a series of masterful experiments, known as the "oil drop" experiment, Millikan and his student Harvey Fletcher not only measured the electron charge, but they also showed that it is the smallest charge in nature and a fundamental unit of all electric charge. They achieved this by exposing a burst of microscopic oil drops to radioactivity. The radioactivity ionized atoms in the oil drops by removing electrons from them. The now charged oil drops, which normally fall to the ground under gravitation, were suspended in the apparatus by an upward electric force countering the gravitational force. From the strength of the electric force required to maintain the suspended drops, one could determine the amount of charge on the drops. Millikan and Fletcher discovered that the amount of charge is always an integral multiple of a unit charge, which they identified as the charge of the electron. For this and later work, Millikan (but not Fletcher) received the Nobel Prize for physics in 1923.

The problems at the forefront of physics research as the world entered its first world war were also changing. They were rendering the hierarchical structure of the European research institute, and of Rowland's ideal academy, less well-suited to advance than was the emerging group-oriented American physics department. It is not just methods, or knowledge, or numbers that ensures the quality of research produced by a community of scientists, but also, and more importantly, the creativity of the people involved and the presence of just the right conditions to foster their work at the forefront of research.

2
American Physics Comes of Age

The outbreak of war in Europe raised important new challenges and new opportunities for physicists on both sides of the conflict. The introduction of poison gas warfare drew chemists into what became known as the chemists' war. But physicists were soon involved as well. Although the United States did not enter the war until 1917, American physicists and chemists began preparations for war as soon as the first shots were fired. This brought physics into a closer alliance with industry, government, and the military. The physicists' strategies for promoting their profession during the war carried over into the postwar decade of the 1920s. These alliances and strategies, forged in the cauldron of war by adept scientist-managers, helped bring American physics rapidly to the forefront of research by the early 1930s.

A WORLD AT WAR

On June 28, 1914, a Serbian nationalist fired a pistol into the horse carriage of Archduke Franz Ferdinand, killing the designated heir to the

Austro-Hungarian throne and igniting the powder keg of Europe. By August 4, Germany and Austria-Hungary, the Central Powers, were at war with Britain, France, and Russia, the Triple Entente or Allies. As the European nations bogged down in trench warfare at the end of the year, across the Atlantic Americans debated what to do.

Many Americans still had family ties to Europe. Most supported President Woodrow Wilson's declaration of American neutrality. Many also ascribed to the progressive view of war that had brought Wilson to the presidency: that war represented a reversion to humanity's barbaric past. War was unworthy of modern civilized nations enlightened by the force of reason, especially the force of scientific reason. For many educated people, the application of science to warfare was simply unthinkable. Germany's introduction of poison gas created by German chemists and the use of science for the technologies of submarines, airplanes, mines, and enhanced artillery were regarded by many as moral violations of natural law. The leader of the progressive-minded Society for Ethical Culture in New York prophesied: "The time will come when the scientist will be considered and will consider himself a disgrace to the human race who prostitutes his knowledge of Nature's forces for the destruction of his fellow men."[1]

Despite Henry Rowland's similar outlook, most American science leaders did not seem bothered by such thoughts, nor were they content to remain neutral. They saw the war as a chance for the country, if it entered the war, to become an influential world power; and they saw it as a chance for American science to rise with the nation to higher status and influence on the domestic and international stages. Leading figures such as Thomas Edison and astrophysicist George Ellery Hale lobbied the president to begin promoting war-related research as a contribution to "preparedness" for the nation's likely entry into the war on the Allied side. The United States finally entered the war on April 6, 1917, with the blessing of most of the nation's scientists. The chemistry department at Harvard "became practically a section of the War Department," writes Harvard's historian. Hale declared the war "the greatest chance we ever had to advance research in America."[2]

By the end of the war two and a half years later, the result was a tighter integration of fundamental research into the nation's economic and cultural affairs and a positioning of physical science, especially physics, for acceleration onto the forefront of research in little over a decade. The war also brought forth a characteristic feature and driving force of American success in physics during most of the century: the appearance of a few powerful scientist-administrators, "scholar-politicians," as one historian calls them, who made it all happen.[3] Prepared for such a role through their experience as leaders of expensive, large-scale research projects, these managerial physicists applied their unique abilities in skilled diplomatic maneuvering and good business sense within and across the boundaries of physics and among their partners in government, business, the military, and philanthropy. It was a role that George Ellery Hale perfected and played to great advantage before and after the outbreak of war in 1914.

A man of driven personality and unbounded ambition, George Ellery Hale was born into a wealthy family in the upscale Kenwood section of Chicago in 1868. After attending private schools in Chicago, he graduated from MIT in 1890 and performed research at the Harvard College Observatory. He received his doctorate in physics in 1894 at the University of Berlin, where he attended lectures by the future quantum theorist Max Planck. Hale's interest in astronomy at an early age arose from his father, an amateur astronomer who built an astronomical observatory equipped with a 12-inch reflecting telescope at the family home in Kenwood. In 1890, the father appointed his son director of what he called the Kenwood Astrophysical Observatory. The combination of astronomy with physics to form the new discipline of astrophysics was at that time still a novelty.

While at MIT, Hale invented the so-called spectroheliograph, a device for studying the sun's physical properties through analyses of the spectrum of light it emitted. Hale used the reflecting telescope together with his heliograph, which now incorporated Rowland's diffraction grating for creating the spectra, in a careful study of sunspots. His work led to the discovery of their unusual magnetic properties. In 1892, Hale became professor of physics at the University of Chicago. Following his European studies, he founded and for many years edited the *Astrophysical Journal*, still a leading journal in its field to this day.[4] Much as did the *Physical Review* for physicists, Hale's journal helped promote the professionalization of astrophysics, as noted earlier, one of the first of the "hyphenated" physics disciplines to emerge during the century.

The year 1897 marked the beginning of Hale's career not as an astrophysicist per se but, presaging the big accelerator physicists of later decades, as the builder of big telescopes funded by big budgets provided by wealthy philanthropists. The rise of the oil, steel, and railroad industries over the past decades had produced a class of immensely wealthy entrepreneurs such as Andrew Carnegie and John D. Rockefeller, as well as lesser known financiers such as Hale's father and Charles T. Yerkes, the founder of the Chicago mass-transit system. With funds provided by Yerkes, Hale completed the Yerkes Observatory in 1897, housing a 40-inch refracting telescope as well as laboratory space for physics and chemistry research. It is still the largest refracting telescope used in research. With the backing of the Carnegie Institution in Washington, D.C., which administered the Carnegie endowment, Hale began work in 1904 on the largest telescope at the time, the 60-inch reflecting telescope built on Mt. Wilson near Pasadena in Southern California. It used a mirror ground from a piece of glass that had been cast in 1896 as a gift to Hale from his father. This telescope was followed at Mt. Wilson in 1917 by the 100-inch Hooker Telescope, funded by John D. Hooker and the Carnegie Institution. It was with these telescopes during the 1920s that Edwin Hubble discovered not only the existence of other galaxies beyond our Milky Way galaxy, but also the surprising observation that the galaxies are moving away from Earth and each other, indicating that the universe is expanding. Using these telescopes, Albert Michelson, who had since joined Hale at Mt. Wilson, made the first measurements of the diameters of stars. Hale died in 1938, but not before setting in motion the construction of the world's largest reflecting telescope until 1975: the 200-inch Hale Telescope completed on Mt. Palomar in 1948 under Hubble's direction.

Hale's connections with the administrators of wealthy endowments, his managerial abilities, and his driven personality made him the premier managerial scientist-politician of the war years.[5] Hale sought to adapt the large-scale project approach of his big observatories to the promotion of American science as a whole in an effort to promote and integrate science, especially physical research, into the power structure of American society. Hale was, in addition, the foreign secretary of the National Academy of Sciences, an august advisory body to the government, which rendered him well attuned to the status of his nation and his science on the world stage.

With the outbreak of war in Europe, the mobilization of physics in preparation of America's possible entry into the war was a natural target for Hale's ambitions. Except for the Army Signal Corps and the Naval Observatory, scientific research in the military was practically nonexistent. Moreover, there was no federal agency capable of enlivening it or of overseeing the mobilization of science in public and private venues. Because of the nation's official neutrality, nothing could be done even to prepare for war until May 7, 1915. On that date, to the nation's horror, a German submarine sank the unarmed British passenger ship *Lusitania* with the loss of nearly 2,000 civilian lives, among them 128 Americans. Preparedness now became official policy, opening the door to the centralized administration of the nation's science for the future war effort. The navy established a Naval Consulting Board chaired by the nation's foremost inventor, Thomas Edison. The board consisted mainly of inventors and engineers, to the exclusion of academic "pure" scientists. Leaving the purity of pure science far behind when it came to military matters (unlike political matters), the academic Hale swung into action, inducing the National Academy to offer its scientific services directly to the president in the event of war. Upon Wilson's reluctant agreement, Hale and colleagues, aided by a presidential mandate, established the National Research Council (NRC) in June 1916 as a subunit of the National Academy. Its purpose was to mobilize and coordinate scientific research for "national security and welfare."[6] It was to encompass scientists of all types and venues, working not just on military applications but on all forms of pure and applied research. Hale named himself permanent chair of the NRC, and he named his like-minded Chicago colleague Robert A. Millikan executive officer and chair of the NRC's physics committee.

Philanthropic foundations and industrial research laboratories hastened to join the new Research Council, while MIT and the Throop Polytechnic Institute in Pasadena, of which Hale was a board member, geared up for war research. The varied scientists of the Research Council joined together in Hale's vision of promoting American physical science through dedication to the common cause. At the center of the NRC stood the physics committee. As physicists from Harvard, Cornell, Chicago, and elsewhere joined the common cause, the insistence on "disinterested pure science" quickly succumbed to the war effort. John J. Carty of AT&T Research declared an end to the "conflict of pure and applied science," according to the minutes of the NRC's first meeting, "pointing out that they do not differ in kind but merely in the objects to be accomplished."[7]

But aspects of Rowland's ideal of pure science lived on. Hale and Millikan conceived of the NRC as an elite organization of private academics, although now with the public goal of organizing and promoting programs in military research. Still in sympathy with Rowland's aversion to any political or social involvement, the science managers sought to render the NRC independent of any government influence or direction by relying solely on the august National Academy of Sciences for authority and on Carnegie, Rockefeller, and similar private foundations for funding. The political naïveté of their position became evident after the United States entered the war in 1917. As a civilian organization, the NRC could not sponsor any military research without military approval. Military authorities were naturally reluctant to accept the directives of an outside agency funded by private philanthropists, and the military was prohibited from providing funds to a civilian organization. In addition, military officers worried that civilian pure scientists would stray from military goals by pursuing research "for the future benefit of the human race."[8]

General George O. Squier of the Army Signal Corps had received a doctorate in physics from Johns Hopkins University. He was eager to tap the expertise offered by Millikan's NRC physics committee for research on improving fighter aviation, then under the Signal Corps. Squier hit upon a simple solution that eventually became widespread. The solution was to absorb Millikan and his physicists directly into the army as commissioned officers. Millikan readily accepted a commission as a major in the army reserve, as did Theodore Lyman and Augustus Trowbridge at Harvard, Albert Michelson at Chicago, and Charles Mendenhall at the University of Wisconsin. Even the president of the National Academy of Sciences received an army commission. As A. Hunter Dupree writes in his classic history, "With this beginning, Squier put virtually the whole physics committee into uniform and hence under orders. By capturing the executive officer himself, he acquired a certain military control over the whole NRC."[9]

However eager they were to avoid government political influence, Millikan and his colleagues expressed no regrets at their sudden subordination to military authority. Millikan's only concern—which proved groundless—was that, as an army officer, he might be hindered from working on naval research. When Millikan's physics unit was transferred from the Signal Corps to the newly established Air Corps, Millikan simply exchanged his old insignia for a set of wings. Even though independent pure research and the training of students suffered as a result, the physicists eagerly embraced their new roles as military researchers, apparently because, as Millikan recalled about the NRC, it promoted the status of his profession and supported, he wrote, "what I regarded as America's responsibility in the war."[10] The ideological instrument of independent pure science was, under the circumstances, not needed. It was a pattern that repeated itself in remarkably similar terms during and after the next world war.

In addition to aviation, two new areas of research especially required the physicists' services. One concerned the detection of hidden enemy artillery batteries by sound ranging. The other entailed the detection of enemy submarines, also by sound, an especially urgent need after Germany unleashed devastating unlimited U-boat warfare.[11] Millikan and his committee, working with British and French physicists, induced the navy to establish a submarine detection center at New London, Connecticut, under the direction of Willis Whitney, still director of General Electric (GE) Research. By the end of the war, the center employed thirty-two physicists, many drawn from the best physics schools in the nation: Yale, Chicago, Cornell, Harvard, and Wisconsin. Hale noted that the war had "forced science to the front" (with plenty of help from Hale), and not only in government and industrial laboratories. By the end of the war, the army and navy (the only military branches at that time) were sponsoring research for the first time in academic laboratories at forty American colleges. Along with this support, military security regulations appeared for the first time, apparently again without objection, in the havens of pure research, the university laboratories.[12]

Even though most physicists and chemists immediately returned to civilian life at the end of the war, the pattern was set and the lessons learned for the rapid integration of scientific research into the war effort. Despite the contributions of the physicists to submarine detection, radio, and aviation, World War I had been the chemists' war, the introduction of poison-gas warfare being their most well-known contribution. Nevertheless, chemistry and chemical engineering had shown that future wars would require the utilization of advanced technologies relying on results gained at the very forefront of fundamental research. The technology and the science required could not be achieved by lone inventors or by a few brilliant geniuses, but only by the coordinated efforts of large groups of well-funded researchers working together on a grand scale for a common cause and across disciplines and venues—from industry and government to academic research laboratories. After Hale and colleagues had shown themselves to be able partners in the common cause, fundamental physics could no longer be ignored by any sector of society, even in peacetime. Those lessons came in handy the next time the nation found itself at war. And, the next time, a new group of scientist-politicians was ready to guide physics once again into a secure position within the military chain of command.

But as the nation entered the postwar world of the 1920s, the status of American physics as an equal partner was hampered by its obvious second-rate position on the international stage behind Europe, and especially behind the nation's number-one wartime enemy—Germany.

SETTING THE STAGE

At the stroke of 11:00 a.m. on November 11, 1918, the guns fell silent across the Western Front. The peeling of church bells rang out the Armistice across the United States. The war to end all wars had suddenly reached an end. Within a month, now Lt. Col. Millikan dutifully unpinned his wings and exchanged his army uniform for the white shirt, tie, and suit of a physics professor. The shooting war was over with a victory for the Allies, but, for him, the battle to bring American physics to the top of world research, then occupied by Europeans, had only just begun.

American participation in the war had been so brief that industrial production and the work of the NRC were still accelerating when the war abruptly ended. Industrial production continued at a high rate after the war as the nation became more business friendly under the three republican presidents who succeeded Wilson through 1933: Harding, Coolidge, and Hoover. Following industry's positive experiences with physics and chemistry during the war, research became an integral part of the postwar corporate economy. According to one count, the number of industrial laboratories grew from 300 in 1920 to 1,625 in 1930, employing some 34,000 researchers.[13] Such an expansion required an increased supply of new graduates and doctorates in physics and chemistry, which required in turn an increase in the number of academic research jobs for the physics professors to teach them. By 1920 a doctorate had become almost a prerequisite for entering the field. According to data compiled by Margaret Rossiter, in 1921 roughly 80 percent of all male physicists (864) but only 63 percent (15) of the 24 female physicists held a doctorate in physics. As a result of the increased demand, the annual U.S. production of physics PhDs nearly doubled during the 1920s to 90 in 1930, then it doubled again during the 1930s, reaching 181 per year by 1940.[14] As shown in Table 1 in the Appendix, by 1938 the number of male physicists had more than doubled from 1921, while the number of female physicists, though still relatively low, had nearly tripled to sixty-three.

Although the number of female physicists badly lagged their male counterparts (to be discussed further in this chapter), the overall increase of interest in physics was reflected not only in the quantity of physicists but also in the quality of their work, as measured on the scale of international recognition and influence. In addition to the demands of industrial laboratories, the main driving force behind this growth was a deliberate strategy pursued from the very start of the postwar period by the NRC and its physics committee under the able and ambitious direction of Professor Robert A. Millikan. "In a few years," Millikan rightly prophesied in 1919, "we shall be in a new place as a scientific nation and shall see men coming from the ends of the earth to catch the inspiration of our leaders and to share in the results which have come from our developments in science."[15]

Born in 1868 in Morrison, Illinois, Robert A. Millikan was descended from a New England family that traced its roots to colonial times. His grandparents had settled in the Midwest as pioneers, and his father was a minister. Millikan grew up on an Iowa farm. He attended Oberlin College and Columbia University where he received his doctorate in physics. After a year of postdoctoral study in Germany, he settled at the University of Chicago where he did his most important experimental work. Following his famous "oil-drop experiment," Millikan undertook a series of experiments in which he hoped to disprove Einstein's revolutionary hypothesis of 1905 that, under some circumstances, light behaves as if it consists not of waves as was commonly held, but of minuscule bundles or quantities— quanta—of energy. Einstein had argued that these light quanta (singular: quantum) could account for the puzzling photoelectric effect, the ejection of electrons from metals when they are irradiated with light of sufficiently high frequencies. Einstein predicted that the energies of the ejected electrons were directly related to the energies of the individual light quanta. Instead of disproving this relationship, to his surprise Millikan confirmed it in 1916 and, along with it, Einstein's explanation of the photoelectric effect based upon the quantum hypothesis. These works led to Nobel Prizes in physics for Einstein in 1922 (for 1921) and, as noted earlier, for Millikan in 1923.[16]

In 1921 Millikan left Chicago for Pasadena as the board chairman of what had been the old Throop Institute but what was now the newly named California Institute of Technology, Caltech for short. Millikan's managerial colleague George Ellery Hale had served since 1904 as a trustee of Throop and had recently obtained a private grant for founding the Norman Bridge Laboratory for Physics at the new Caltech. Millikan could not resist the opportunity to build up the new laboratory as its director and the new institution as its board chairman to first-class status at what he called "the westernmost outpost of Nordic Civilization."[17]

As with the nation's physics as a whole, Millikan's ambitious aim for the privately funded, all-male institution was no less than to create "as outstanding a center of research *in the field of all the sciences* as the world possesses" (his emphasis).[18] Once again seeking to avoid the taint of "political trickery," he refused to accept any public funds for his institution. Equally at home in the boardroom and the laboratory, Millikan saw himself as what we would call the CEO of Caltech. In the business-friendly environment of the 1920s, he too epitomized the managerial approach to the administration of American science that helped make it what it became. But some were not impressed. Physicist Hans Bethe later recalled of the man, "Millikan was very much convinced of his own importance. I thought he was very pompous."[19]

A number of broader developments, both in the United States and in Europe, were coming together after the war to enable the success that Millikan and a few other physicist-politicians managed to achieve in so short a time. These included, on the domestic front, a postwar increase in the number of students seeking careers in science in concert with a rapid growth in the total number of students. In 1900 there were about 82,000 college students in the United States; by 1930 there were 1.1 million, amounting to 7.2 percent of the college-age population. Many had received a greatly improved preparation in high-school mathematics, chemistry, and physics.[20] Other favorable factors for American physics included a fading of task-oriented military research (although the navy opened the Naval Research Laboratory in 1923 at Edison's urging); a new concern in business and philanthropy for the support of academic pure research; and, especially, new breakthroughs in relativity theory and quantum mechanics occurring abroad that opened new fields of research into understanding old and new phenomena.[21]

In 1919 British astronomers astonished the world by confirming Einstein's prediction, based on his general theory of relativity, that the paths of starlight passing close to the sun during a total solar eclipse would appear to be bent as they passed the sun. Overnight, Einstein became an international celebrity. According to Einstein's theory, the massive sun slightly curves the space around it, thus causing the beams of starlight to bend toward the sun as they pass near it on their way to Earth. General relativity soon became an important component of astrophysical research, especially after Edwin Hubble's startling discoveries at the Mt. Wilson observatory regarding the existence of other galaxies and the expanding universe.

Six years after Einstein's triumph with the confirmation of his theory in 1919, German and other European theorists, including Heisenberg, Schrödinger, Pauli, Bohr, Dirac, and Born, astonished the world once again with the second of the two great revolutions in twentieth-century physics—the breakthrough to quantum mechanics. The new theoretical physics of the atom has proved indispensable ever since in understanding the behavior of atoms, molecules, solids, stars, and subatomic particles.

PUSHING THE PEAKS HIGHER

Some of the nation's leading physicists populated Millikan's postwar NRC physics committee. In addition to astrophysicist George Ellery Hale, they included experimentalists Percy W. Bridgman and George Washington Pierce at Harvard, Harrison M. Randall at Michigan, and the brothers Karl T. Compton at Princeton and Arthur H. Compton at Washington University, St. Louis, and the University of Chicago (beginning in 1923).

Even as nontheorists, they were fully aware of the latest theories coming out of Europe, and all were aware that the United States would have to compete directly with Germany and other European nations in the field of theoretical physics, in addition to experimental research, if it were to compete in the international arena and maintain the vitality of its research.

Since the early nineteenth century, physics had been defined in most nations as experimental physics—the acquisition and analysis of laboratory data about the workings of physical phenomena. Theoretical physics, the mathematical investigation and expression of the laws of nature as epitomized by James Clerk Maxwell's theory of electromagnetism or by the statistical mechanics of gases developed by Maxwell, Boltzmann, Gibbs, and others, was usually considered inferior to the direct encounter with nature in the laboratory. Even in Germany, with its strong traditions of mathematics and, now, theoretical physics, the discipline was held in low regard until the experimental confirmation of Einstein's general theory of relativity and the breakthrough to quantum mechanics.[22] Einstein's well-publicized visit to the United States in 1921, during which Princeton University conferred on him an honorary doctorate, only highlighted American deficiency in theoretical work.

By the 1920s experimental research in the United States was focusing more directly upon such highly technical areas as electromagnetic fields, atomic and molecular spectroscopy, and the atomic structure of matter. Some of this work exerted a direct impact on electrical engineering and the physics of solids and gases, of interest to industry. With the new developments in quantum theory, these areas of research were requiring sophisticated mathematical and analytical skills beyond the means of many experimentalists. Theoretical physicists adept in the new theories yet closely attuned to experimental research were becoming essential to the future progress and competitive ability of American physics. In 1919 Harvard physics professor Percy W. Bridgman induced his university to begin offering graduate courses in theoretical physics. He wrote to Edwin C. Kemble, newly recruited to teach the courses: "If we can get the courses well given, it ought to put Harvard pretty near the top in this country. What is more, it is a good beginning to putting this country on the map in Theoretical Physics."[23]

Drawing upon the NRC's wartime policies and Rowland's pure-science elitism, Bridgman, Millikan, and the other NRC experimentalists introduced a strategy for institution building aimed at catching up with the Europeans as quickly as possible. They called it "making the peaks higher."[24] The strategy was to utilize the generosity of the private Rockefeller and Carnegie endowments to push a few of the best American research universities to even greater heights through targeted research grants, rather than to bring a large number of the nation's universities up to a somewhat lower level of excellence. As in Rowland's day, this left the other colleges and universities to languish in the lower tier. Among the peak universities chosen were, of course, those represented on the NRC physics committee: Caltech, Harvard, Chicago, Michigan, Princeton, MIT, and a few others. The formula succeeded. By the early 1930s these institutions were among the top twenty research universities in the country, and they were producing work that drew increasing international attention. Throughout the 1920s these institutions housed the majority of the nation's academic physicists, produced more than 75 percent of the papers published in the *Physical Review*, and produced 90 percent of American doctorates in physics.[25] They are still among the nation's top research universities in physics and other sciences.

In order to cultivate a new generation of physicists equipped to work at the rarefied level of the peak universities and especially in theoretical physics, as early as 1919 Millikan and his managers began funneling Rockefeller funds into postdoctoral research fellowships granted to roughly the top 15 percent of graduates in physics (followed later by chemistry, medicine, and biology). The fellows could work at a university of their choosing or perform research abroad under the auspices of the Rockefeller Foundation's International Education Board (IEB). The aim was to spark research in the departments they visited, to enable young physicists to learn the latest research abroad, and to prepare the fellows for careers devoted to research and to the training of the next generation of first-class physicists. Most fellows chose to work at the select American universities or at one of the leading European research institutes. The Rockefeller Foundation provided the funding; the NRC made the selections and administered the program. At the top of the NRC's wish list for fellows were those eager to learn the new quantum mechanics at its source in Europe or to train under prominent domestic researchers or European theorists invited to teach the new physics in the United States as visiting or permanent professors. The effects were soon evident. In 1925 one-eighth of the papers by American authors published in the *Physical Review* acknowledged support from a NRC fellowship.[26]

Millikan's and Hale's Caltech was naturally at the center of this strategy, hosting fully half of the NRC fellows and receiving nearly every visiting European theorist for periods from a brief sojourn to a year. Among the visitors passing through Pasadena during the late 1920s were four of the founders of quantum mechanics: Erwin Schrödinger, Werner Heisenberg, Paul Dirac, and Niels Bohr. Heisenberg arrived in Pasadena from the University of Chicago, where his lectures on quantum mechanics during the spring of 1929 became a popular textbook. Together with Dirac's lectures, *Principles of Quantum Mechanics* (1930), they served as the first introductions to the new physics for many American physicists.[27]

In 1925 Millikan brought Swiss cosmologist Fritz Zwicky to Caltech as a permanent faculty member through the IEB. With the Mt. Wilson observatory at hand and the arrival in Pasadena of general relativity theorist Richard Tolman, Caltech became a leading center for astrophysics research. For quantum atomic theory, German physicist Paul Epstein, who had worked with Arnold Sommerfeld in Munich, accepted a permanent position at Caltech in 1921. But Millikan turned down the opportunity to hire another outstanding mathematical physicist who was Jewish because he believed that Caltech could not tolerate a second Jewish faculty member. Nor, he believed, would a new Jewish faculty member be compatible with the overwhelmingly Anglo-Saxon population of Southern California, some of whose members provided Caltech's financial support.[28] Apparently, however, fame trumped prejudice, for Millikan persuaded Einstein to return annually to Caltech as a visiting professor from 1930 to 1932. His visits showcased the international prominence that Caltech had achieved by then in physics, and they served, more practically, as a good argument for gaining private donations.

In 1927 Harrison Randall and the University of Michigan hosted the first of the annual Ann Arbor summer schools for theoretical physics. During these events, which ran through the late 1930s, American students and faculty gathered to learn the latest theoretical and experimental research from top foreign and domestic physicists. After speaking at the first summer school, Dutch theorists George Uhlenbeck and Samuel Goudsmit joined the faculty of the University of Michigan in 1927. German quantum theorist Alfred Landé landed at Ohio State in 1931. During an invited sojourn at MIT in the winter of 1925–1926, German physicist Max Born, one of the founders of quantum mechanics, reported to his dean at the University of Göttingen that he had already received two job offers. "While in the United States the experimental physics is in high blossom," he explained, "the theoretical physics is entirely undeveloped. Europe's lead must be recognized for the moment; but at the same time the ambition of the Americans is directed toward the goal of filling this gap, initially by attracting European assistance."[29]

Although experimentalists far outnumbered theorists in the United States, all eagerly utilized the new physics in a variety of areas. European physicists began to take note. Even before quantum mechanics, John H. Van Vleck's quantum theory of the structure of the helium atom and Kemble's efforts to unravel the infrared band spectra of diatomic molecules had drawn the interest of German quantum mechanicians. As Europeans in Bohr's circle considered whether or not the puzzling light quanta really do exist, A. H. Compton's confirmation in 1922 of their existence in his classic experiment at Washington University, St. Louis, on the scattering of X-rays by free electrons immediately caught their attention. His subsequent experiment with Alfred W. Simon on the directed scattering of light quanta, published in 1925 in the *Physical Review*, refuted a controversial statistical theory put forth in 1924 by Bohr, H. A. Kramers, and Harvard physicist John Slater in an effort to avoid the problematic existence of light quanta. Suddenly, references to papers published in the *Physical Review* began to appear in the papers and private letters of Pauli, Heisenberg, and Bohr.[30] In 1927 Compton received the Nobel Prize for his work. Following Heisenberg's breakthrough to the matrix version of quantum mechanics in 1925 and Schrödinger's alternative wave mechanics in 1926, American physicists jumped into quantum research, producing valuable applications of the new physics and becoming an audience the Europeans could not longer ignore. Perhaps because he was in competition with Heisenberg over which of the two versions of quantum mechanics was preferable (they were soon proved equivalent), Schrödinger thought it valuable in 1926 to publish an English account of his new theory in the *Physical Review* for American readers. His widely read paper, "An Undulatory

Theory of the Mechanics of Atoms and Molecules," appeared in volume 28 of the journal.[31]

Altogether, the components of the American strategy—inviting foreign specialists as lecturers and professors, funding the best American post-doctoral researchers to work at the best American universities, sending top young physicists abroad to learn the new physics at the source, and funneling philanthropic funds into building up research programs at the best universities to even higher quality—began to have an accelerating effect. "Quite suddenly," writes historian Spencer Weart, "the spirit of European theoretical physics flowed into America."[32] According to data compiled by Yves Gingras, beginning in 1926 the annual number of physics papers published in the United States rose sharply from about 350 per year to 1,200 per year by 1938. During the same period, the annual number of such papers decreased in the United Kingdom from 800 to under 700, while those in Germany increased briefly from about 350 in 1926 to 450 in 1931, then sank back to about 325 in 1938 as the Nazi regime tightened its grip on the country.[33] As Americans began to compete more directly with Europeans at the forefront of research, the *Physical Review* began to appear biweekly in 1929 and with a new section of "Letters to the Editor" that quickly announced new research results ahead of foreign competitors. In 1958 the section became an essential independent publication, *Physical Review Letters*.

According to one assessment, by 1930 the theoretical physics faculties at four universities—Caltech, Berkeley, Chicago, Michigan, and Princeton—were comparable in quality to the great European institutes, and six other faculties were close behind: Harvard, Columbia, Johns Hopkins, MIT, Cornell, and Wisconsin.[34] A year later, the tide had turned. "In that year 1931," writes Weart, "the *Physical Review*, for the first time, was cited more often in the physics literature than its chief rival, the German *Zeitschrift für Physik*."[35]

ENCOUNTERING HURDLES

With the advent of general relativity theory and quantum mechanics, new frontiers of research were opening in physics during the 1920s as perhaps never before, and American physicists seized the opportunities available to them. Nevertheless, despite their brilliant strategy for success and the large numbers of new, first-rate students available to help achieve that success, an ugly facet of American academic policy worked against this achievement. As in other nations at the time, discrimination against women and ethnic, racial, and religious minorities hindered and, in many cases, prevented large numbers of potential future scientists from contributing to the nation's success and making it even grander. The discrimination against these groups was even greater in an era when pushing the peaks higher also meant pushing the barriers to entrance higher as well. Millikan's all-male college at the edge of "Nordic civilization," yet at the center of the drive for excellence, certainly did not help matters.

Discrimination in the United States and in science in particular has a long and complicated history. Margaret Rossiter's now classic two-volume study, *Women Scientists in America*, offers a wealth of information and insight regarding the "struggles and strategies" of women scientists through 1972.[36] The number of women physicists in the United States during the 1920s and 1930s was far below the number of their male counterparts. This held, of course, for other sciences as well. Table 1 in the Appendix summarizes some of Rossiter's data obtained from the annual listings in *American Men of Science* for the years 1921 and 1938. Although only a fraction of the total number of physicists and scientists, presumably the most successful, were selected for this publication, the data do suggest several causes for the general trends. First, it is remarkable that in both sample years physics drew only single-digit percentages of the total numbers of men and women scientists. The percentages, in fact, declined for both genders from 1921 to 1938, even though the number of physicists more than doubled. This may have been because by 1938 physics was no longer a newly developing field in the United States, while more women had since found other fields more accommodating. According to Rossiter's data for the other sciences, in both 1921 and 1938 chemistry, the medical sciences, and engineering were far more popular with men, while botany, zoology, and psychology were most popular with women scientists. A recent study has shown that discouraging stereotypes as well as subtle bias and not-so-subtle discrimination can have significant impacts on women's choices for future careers today, and surely did so then.[37]

Although World War I had brought many women scientists into new jobs in industry, government, and academe, they lost many of those jobs after the war as the male scientists returned from their war work with the

nearly all-male NRC and military branches. Despite the momentum generated by the passage of universal suffrage in 1919, the prewar segregation
of women into "women's work" in science returned after the war. Many
men considered women mentally unsuited for scientific thinking, a view
apparently held even by the *New York Times*. In an unsigned editorial appearing on June 4, 1921, the paper stated: "That women can be efficient in
laboratories . . . needs no proof at this late day. It is still true, however,
that the majority of women are still to develop either the scientific or the
mechanical mind." The *Times* explained that it is not that women are
inferior to men, "but that more men than women have latent capabilities
in those directions . . . [capabilities] of viewing facts abstractly rather than
relationally, without overestimating them because they harmonize with
previously accepted theories or justify established tastes and proprieties,
and without hating and rejecting them because they have the opposite
tendencies."[38]

The numbers of PhDs by gender reveal more of the story (see Table 1
in the Appendix). Prewar discrimination in doctorates was less virulent
than in employment. As noted earlier, in 1921 roughly 63 percent of women
physicists held PhDs, while 80 percent of men physicists did so. By 1938 the
percentages, if not the numbers, were nearly equal, 73 percent of women,
75 percent of men. But when one looks at the numbers of *all* science PhDs
in *American Men of Science* by gender, the picture is just the reverse! In
1921, 72 percent of all women scientists held doctorates, compared with
only 58 percent of all male scientists. In 1938 it was 83 percent of the
women scientists and 70 percent of the men.

What was different about physics? And why were women scientists as
a whole more likely to hold doctorates? The answer to the first question
apparently lies in the circumstance that in 1921 a doctorate was strongly
preferred but not required for academic physics. In a study of the twenty-
four women physicists listed in the first three editions of *American Men
Science* (1906, 1910, 1920) and still employed in 1921, Rossiter found that
nineteen of the twenty-four women were employed at ten women's colleges. (However, only twenty-one appeared in *AMS* in 1921.) The remain-
der included high-school teachers, a high-school science director, and
physicists at the National Bureau of Standards and at the H.C. Keith Company. Fifteen of the twenty-four women physicists were listed as professors,

of whom eleven held doctorates. Interestingly, four of the other nine women also held doctorates, including the high-school science director—a likely indicator of the segregation into "women's work." Only two of the doctorates were awarded by German universities. The rest were awarded by Cornell (three); Bryn Mawr, Chicago, Johns Hopkins, and Pennsylvania (two each); and Clark and Tulane (one each).[39] By 1938, for those listed in *American Men of Science* a doctorate was an even greater requirement of both genders in physics, hence the near equality of women and men by percentage.

Why were women scientists overall more likely than men scientists to hold doctorates throughout this period? The answer apparently lies in what Rossiter calls the "Madame Curie strategy." Many women perceived that to be accepted as equal to men "in a man's occupation" they had to be demonstrably better than men, much as was Madame Curie, the recipient of two Nobel Prizes (for physics and chemistry). And, like Madame Curie, who had to overcome incredible discrimination and other obstacles, many women believed they had to face the discriminatory world of science with a stoic acceptance in the hope that, if they could survive, they would ease the path for those who came after them. As a result, many became, writes Rossiter, "the extraordinarily over-qualified, uncomplaining, and utterly self-sacrificing paragons that abound in the obituaries of women scientists."[40]

Ethnic, racial, and religious minorities faced similar discrimination. But its manifestations were tied more closely to the waves of immigration that the United States, the "nation of immigrants," has experienced since its founding. On the whole, as each new immigrant group arrived and established itself, it tended initially to resist the next wave of immigrants. In many instances, new immigrants were perceived as competitors for jobs, including future professional jobs.

Although there are many similarities, each ethnic group has experienced its own unique circumstances. While African-Americans encountered blatant racism, segregation, and the after-effects of long-term slavery, Asians were generally excluded from migration to the United States until after the World War II, except when they were needed for the railroad and other labor-intensive projects. Aside from the immoral and impractical aspects of these practices, it is sad to think of the potential lost contributions these and other groups could have made to American society and science, and that, by their underrepresentation, they are failing to make even today. The widespread imposition of official quotas against the enrollment of Jewish students at many American universities is one of the most appalling episodes of twentieth-century American academic history. This lasted officially from the early 1920s until 1948, and perhaps unofficially thereafter in some quarters.

Anti-Semitism in the United States also has a long history, as does its expression in the academic realm. One of the factors bringing such underlying prejudices into the open included the record number of persecuted Jewish immigrants arriving in the United States during the prewar years at a time when earlier arrivals were just rising into the middle class through education and business. The Great War itself set off fanatic waves of superpatriotic and nativist demands for total assimilation to "American" cultural ideals and war aims. Anti-immigrant violence mounted during the war, while racism and racial anti-Semitism spread throughout American culture after the war. "Alien races," it was argued, could not assimilate into American culture and must therefore be excluded, by law or by force if necessary.[41] In 1921 and 1924, the United States introduced immigration quotas for the first time. The quotas were established by country of origin and were directed mainly against immigrants from southern and eastern Europe, that is, mainly against Jewish, Italian, and Slavic immigrants.[42]

In parallel with immigration quotas, colleges and universities began to impose quotas on the admission of Jewish students. For reasons of tradition and discrimination, the first-generation of the newly immigrated Jewish families were the most academically mobile of any ethnic group. Nearly two-thirds of Jewish young people went on to college. Some universities feared being "overrun" by Jewish students. As ethnic historian Marcia Graham Synnott writes, "Their conspicuous academic success made Jews potential competitors of the white, native-born, Protestant middle class."[43] Many universities, especially the elite research universities that were bringing American science to international competitive status, regarded themselves as guardians of professional quality and as gateways to professional careers. Thus, they regarded it as their duty to society to maintain the quality, uniformity, and social cohesion of the educated elite that would run the future American culture and economy, science and government. At the same time, according to Synnott, most faculties generally adhered to the commonly held notion of racial superiority and were fearful that large numbers of graduating Jews would not fit into "Nordic society." As a result of all this, most universities began introducing quotas designed to preserve the ethnic makeup of their student population and of the future professional class. And most faculties saw to it that at most only a handful of faculty members, including physicists, were of Jewish descent.[44]

According to data compiled by Synnott, Jewish enrollments rose sharply after the war as immigrant children reached college age. In 1922 Jewish students constituted 21.5 percent of Harvard's freshman class, over double the percentage for the student body of 1918–1919 (10.0 percent). At Yale, the percentage of Jewish students jumped from 7.6 percent in 1918–1919 to 13.3 percent during the 1920s; for Princeton, the percentage increased from 2.6 percent to 3.9 percent during the same period.[45]

The largest Jewish population was in New York City, and its colleges and universities enrolled the largest numbers of Jewish students. In 1919 Columbia and New York universities became the first of a group of universities concerned about Jewish enrollments to introduce special testing to identify and reduce the number of Jewish admissions. During the next three years, Jewish enrollment in the freshman year at Columbia dropped from about 33 percent to 20 percent. The Jewish representation stayed at about that level for the rest of the decade, dropping further during the 1930s to only about 15 percent in 1934–1935. However, at New York University any quotas imposed during the 1920s, for which, unfortunately, no data are available, had disappeared by 1934, when the Jewish representation was nearly the same as it had been in 1918–1919, around 45 percent.

Harvard president A. Lawrence Lowell, a leader of the national movement for immigration quotas, was especially concerned about the rapid increase each year in the number of Jewish students enrolling in Harvard's freshman class.[46] In an attempt to discourage Jewish applicants, Lowell announced in an interview published in 1922 on the front page of the *New York Times* that henceforth admissions of Jewish students would be limited to 15 percent, a figure that, he believed, "might help them." He explained that the presence of greater numbers of Jewish students on campus was causing an increase in "anti-Semitic feeling among the students."[47] It was the classic shifting of blame to the victims for their treatment. Although the number of Jewish freshmen jumped another 6 percent by 1925, a year later only 15 percent of the Harvard freshman class was identified as Jewish, and the number remained in that range for over a decade.

Quotas at other universities soon followed. Those with high initial percentages of Jewish students set quotas at about 10 percent to 15 percent of the student body into the next world war. According to Synnott, in 1934–1935 Jewish students constituted on average 9.13 percent of the nation's student body. Once the most academically mobile group, by 1934 only 56 percent of Jewish college applicants were successful, while 67 percent of Catholics and 77 percent of Protestants were accepted. A parallel pattern of exclusion occurred in jobs and in many of the professions, including physics in academe and industry. Even after the defeat of Germany in World War II, after the horrors of the Holocaust were widely known, the quota system did not officially end until 1948. In July 1945, as the Third Reich lay in ruins, Millikan worried to Richard Tolman about bringing Jewish physicist J. Robert Oppenheimer, the head of the Manhattan Project, back to Caltech after the war had ended. He worried, he wrote, "because of the large percentage of his fellow racists [sic] who are already appearing at the Institute." He felt he could not ignore "my responsibility to the Institute in a world built as it is."[48]

A PRIME EXAMPLE

One of the beneficiaries of the NRC's fellowship strategy was a Harvard University graduate named J. Robert Oppenheimer. The NRC managers hand-picked the young man to help bring American theoretical physics to world-class status. His story illustrates how the NRC's strategy worked and how it helped shape an important figure in American physics during the following decades.[49]

Oppenheimer was born in 1904 in New York City to a wealthy family that had abandoned its German-Jewish heritage and sought assimilation into New York City's upper economic and cultural class. Oppenheimer attended one of the premier private schools in the city, the Ethical Culture School, regarded at the time as a prep school for Harvard and Radcliffe.

Just months before Oppenheimer joined the 21.5 percent of "Jewish" students in Harvard's freshman class of 1922, President Lowell unleashed his campaign for quotas at Harvard with his warning to Jewish students in the *New York Times* about the atmosphere they were likely to encounter. A survey of eighty-three Harvard undergraduates published that September found that roughly half believed a "race limit" was justified at Harvard, while only 34 percent held that it was never justified.[50]

Not a gregarious student in any case, Oppenheimer spent his undergraduate years at Harvard among a small circle of like-minded, mostly non-Jewish students who shared his rarified interests in literature, philosophical debate, and science. He was as prone to spin off short stories and French poetry as he was field equations. Oppenheimer's friends also shared his intense dislike of Harvard, usually expressed in letters as an aversion to what they regarded as the university's shallow academic level and "Puritanical" outlook, although anti-Semitism and political conservatism also played a role. Oppenheimer had headed for Harvard intending to major in chemistry. Having completed the required coursework, including organic chemistry with James B. Conant, Oppenheimer began gravitating toward experimental physics under Bridgman.

Graduating summa cum laude in chemistry in just three years, Oppenheimer began graduate work in experimental physics in Britain at the famous Cavendish Laboratory at Cambridge University. In writing a letter of recommendation to Ernest Rutherford in June 1925, Bridgman thought it necessary to inform Rutherford: "As appears from his name, Oppenheimer is a Jew, but entirely without the usual qualifications of his race."[51]

Oppenheimer soon turned to theoretical physics. Just months before Oppenheimer arrived in Cambridge in September 1925, German physicist Werner Heisenberg, working in Göttingen, had made the initial breakthrough to quantum mechanics, the new physics of the atom. Heisenberg brought the news to Cambridge physicists, among them Paul Dirac, during a lecture visit that summer. By the fall of 1925, Dirac was busily constructing his own version of a complete quantum mechanics on the basis of Heisenberg's ideas. Oppenheimer was enthralled. But at age 22, only two years younger than Dirac, Oppenheimer still did not have the grounding in theoretical research that would enable him to participate. Nevertheless, by the end of the academic year when Heisenberg's mentor, the theorist Max Born, invited Oppenheimer to work with him in Göttingen, the capital of the quantum revolution, the young man had found his future career.

The tall, thin, well-spoken, and always immaculately attired Oppenheimer worked so intensely in Göttingen that he completed his doctorate under Born on an application of quantum mechanics to continuous spectra within a mere four months. He published the work in German in the leading journal for quantum research, the *Zeitschrift für Physik*. He went on to produce works alone or with Born over the following two years on applications of quantum mechanics to such topics as molecular structure, the scattering of alpha rays, and the capture of electrons by atoms. Some were published in German; others appeared in *Nature* and *Physical Review*.[52]

By late 1926 Oppenheimer had already clearly mastered the new quantum mechanics. His ability to hold his own in the capital of quantum mechanics quickly brought him to the attention of American scouts looking for new talent in Göttingen. On a visit to the university town that year, Kemble reported back to Bridgman of their former pupil's stellar performance. Karl T. Compton, Arthur Compton's brother and a member of the NRC fellowship selection committee, wrote from Göttingen in December 1926 to an official of the Rockefeller Foundation, "There are 20 or more Americans here, mostly in Mathematics, Physics, and Medicine. As far as I can learn, [Edward] Condon and a young chap named Oppenheimer are the star performers in physics."[53] At Compton's urging, Oppenheimer applied for an NRC fellowship to work at Caltech with Paul Epstein. But Bridgman wanted Oppenheimer for Harvard. In the end, Oppenheimer split his fellowship during the 1927–1928 academic year between Harvard and Caltech, two of the top four destinations for NRC fellows in physics at that time (the others were Princeton and Chicago).

Fresh from Europe and with a doctorate in theoretical physics from Göttingen, Oppenheimer fit perfectly into Millikan's plans to raise Caltech and American physics to European standards in theoretical research, especially in quantum physics, Millikan's own field of research. By the end of the academic year, Oppenheimer had ten university teaching offers, including ones from Harvard, Caltech, and the University of California at Berkeley. There seemed to be no quotas on job offers to the highly talented young man. But Oppenheimer wanted another tour in

Europe "to try to learn a little physics there," as he put it. He was one of only four American fellows selected that year for an IEB fellowship.[54] After stopping in Ann Arbor for a summer school session in theoretical physics, Oppenheimer headed to Europe in the summer of 1928.

By the time he returned to the United States a year later, Oppenheimer's travels had brought him to nearly all of the European centers for quantum physics: Cambridge, Göttingen, Leiden, Zurich, and Leipzig (but unfortunately not Copenhagen). He had met and worked with many of the leaders of the quantum revolution and had become acquainted not only with them and their physics, but also with many of the young people of his generation of future contributors. He had worked with Dirac and Born, visited with Paul Ehrenfest in Leiden, attended Heisenberg's lectures on his new quantum theory of magnets (ferromagnetism) in Leipzig, and spent a difficult six months with Wolfgang Pauli in Zurich. There he helped trace the seemingly insurmountable problems facing the combination of relativity theory with quantum mechanics in the creation of a relativistic quantum field theory, a theory soon needed for the study of collisions of high-energy elementary particles.

One of the other American fellows dispatched to Europe that year was Isidor I. Rabi, a quantum experimentalist educated at Cornell and Columbia, who was, like Oppenheimer, eager to promote his profession in the new world. The two young men first met in Heisenberg's seminar and became life-long friends. Both also worked in Pauli's institute. Rabi later recalled their embarrassment in Europe: "We were not highly regarded, I must say, nor was there any thought that America would amount to anything as far as physics was concerned . . . We felt very bad about this." The expatriates decided to do something about it. "When we came back, my generation, Condon, [H. P.] Robertson, Oppenheimer, myself, and then people like Van [Vleck], sort of mature enough to get students, we just sort of changed the whole works in this short time." Oppenheimer agreed. One of the ways that the gap with European physics was soon eliminated, he said, "was through Americans who had studied in Europe and came back and decided that there was no reason that things like that shouldn't be going in this country, and went to work on it."[55]

And work on it they did. Rabi returned to Columbia University where, during the 1930s, his work on molecular beams led to the Nobel Prize in 1944. Condon pursued the quantum theory of molecular band spectra at

Princeton and the Westinghouse Research Laboratories. Slater developed the quantum mechanics of complex atomic spectra; Robertson worked on both mathematics and physics at Caltech, then at Princeton; and Van Vleck continued research in quantum theory at Wisconsin and Harvard. In addition, David M. Dennison at Michigan and Robert Mulliken at Chicago, following work by Condon, P. M. Morse, and others, pursued a definitive quantum mechanics of molecular band spectra; Linus Pauling at Caltech began work on the quantum mechanics of the chemical bond, stimulating the field of physical chemistry; and Slater worked in the 1930s on extensions of German theories of ferromagnetism (Heisenberg), the electron theory of metals (Ernst Bloch), and the energy-band theory of conductivity (Rudolf Peierls).[56]

Oppenheimer made a deal with Millikan. Before leaving for Europe, Oppenheimer agreed to take up a split appointment between Berkeley and Caltech upon his return. Oppenheimer liked the vibrancy of Caltech, but he was intrigued by Berkeley, one of the then-mediocre state teaching universities that Millikan and the NRC had targeted for transformation into elite graduate research universities in physics and related sciences. He later recalled: "I visited Berkeley and I thought I'd like to go to Berkeley because it was a desert. There was no theoretical physics and I thought it would be nice to try to start something. I also thought it would be dangerous because I'd be too far out of touch so I kept the connection with Caltech."[57] Since Berkeley needed Oppenheimer more than did Caltech, Millikan placed Oppenheimer at Berkeley as his home institution, but required of him a trimester and a summer each year at Caltech, at least until the mid-1930s. Oppenheimer remained officially at Berkeley until 1947. But there was one more piece to the puzzle of bringing the state university into the research elite.

As the Berkeley physics department hurriedly shifted its resources from undergraduate teaching to graduate research, Raymond T. Birge, in charge of recruitment, pursued the same strategy as Millikan in gaining top faculty: the capture of NRC postdoctoral fellows in physics. In 1928 Birge persuaded Ernest O. Lawrence, a former NRC fellow universally regarded as a future top experimentalist, to leave his new Yale faculty position for a professorship at Berkeley. He promised the young man not only immediate tenure and promotion to associate professor, but also lavish funding and support for his laboratory. He then wrote to Lawrence of his plans for a second open position: "As you may know, we are gradually getting quite a strong department, and are constantly strengthening it. There are two positions open for next year, and besides that being offered to you, we will probably try to get a really able mathematical physicist of the productive kind for the other position."[58]

Oppenheimer was their man: upon his return from Europe, the die was cast for California. The managers of American physics had decided upon Oppenheimer and Lawrence to help maintain and expand the nation's position in the top ranks of research from California. At the same time, the Berkeley physics department had selected Oppenheimer and Lawrence to combine contemporary quantum theory and experiment in a collaborative research program to push the university rapidly to the peaks of American physics.

By the early 1930s it was obvious not just on the basis of the number of physicists or citations of *Physical Review* articles that American physics had "come of age," as Van Vleck put it.[59] American quantum experimental research produced by Rabi, A. H. Compton, Clinton Davisson, Lester Germer, and others was at or near the first ranks of world research, as was theoretical work by Oppenheimer, Van Vleck, Slater, Pauling, Robertson, and others. Sessions on quantum mechanics at the American Physical Society meetings drew crowds of over 400 participants.[60] As Millikan had predicted, foreign physicists were now coming to the United States to learn the latest developments. Slater recalled how impressed he was by the American papers given during an American Physical Society meeting in June 1933: "not so much the excellence of the invited speakers as the fact that the younger American workers on the program gave talks of such high quality on research of such importance that for the first time the European physicists present were here to learn as much as to instruct."[61]

3
Surviving the Depression

Ernest Orlando Lawrence arrived in Berkeley in the fall of 1928. Six months later, he reportedly hit upon the basic idea for an invention that would revolutionize nuclear research, a device for accelerating subatomic particles and smashing them into atomic nuclei, producing untold riches in new knowledge about the nucleus and its constituents. It was soon learned that if these particles could be smashed into matter at even higher energies, they would produce new and exotic elementary particles, providing clues even about the most fundamental forces of nature. Lawrence later called the device he had envisioned a "cyclotron," for the cycles the particles underwent as they accelerated to high speeds. By the end of the year, he and his assistants were hard at work on the difficult task of realizing his dream machine. He wrote to his parents in February 1930: "If the work should pan out the way I hope, it will be by all odds the most important thing I will have done."[1] Indeed it was.

Lawrence's parents were descendants of Norwegian immigrants to the western prairie. Their son was born in Canton, South Dakota, in 1901.

His father was the superintendent of schools in Canton. Like many American youngsters of his era, he grew up tinkering with cars, radios, and farm equipment. These were all skills useful to experimental physics. The tradition of such tinkering practiced by other future experimenters helps to account for the strength of American physics in this research and in the development of new equipment and the electronics to run it.

Lawrence graduated from the University of South Dakota in 1922. He received a master's degree in physics at the University of Minnesota a year later and the doctorate in physics at Yale University in 1925. He was one of the first major physicists of the era educated entirely within the United States, an indication of the growing stature of the nation's discipline. Lawrence remained at Yale as a National Research Council (NRC) fellow for two years and as an assistant professor for one, working on measurements related to Einstein's photoelectric effect. He joined the department of physics at Berkeley in 1928 as an associate professor. Two years later, at the age of 29, he was promoted to full professor, the youngest person to hold that title at the university. It was only after Lawrence traded his investigations on the photoelectric effect for the rising field of nuclear physics that he conceived the idea of the cyclotron.[2]

The birth of the cyclotron was well-prepared for by others before Lawrence. Electrically charged particles are attracted and repelled by electric forces. Some experimenters had already devised methods of using electric forces, or electric potential differences, to accelerate charged ions and subatomic particles, protons and electrons. One way was to accelerate them down a long evacuated tube with constantly alternating electric fields that would give the particles successive "kicks" of attraction then repulsion as they sped past. Another method involved the use of a high-voltage electrostatic charge to accelerate the particles, an idea used in an accelerator named for R. J. Van de Graaff.

Lawrence's brilliant insight was to utilize the unique behavior of moving electric charges in *magnetic* fields. According to a fundamental law of physics, a charge moving through a magnetic field created by magnetic north and south poles brought near each other will experience a force perpendicular to its direction of motion. Constantly subjected to such a force, the path of the charge will bend into a circle. The radius of the circle depends on the speed, charge, and mass of the particle and the strength of

the magnetic field. Lawrence realized that he could accelerate a charged
particle using electric fields as it moved on its magnetically induced circular path if its path traversed through two half circles, or "dees" (since they
are shaped like a D). Each half circle was electrically charged with opposite charges so that the particle is repelled by one dee and attracted to the
second dee. After the particle accelerates across the gap between the two
dees, it nearly completes a half circle through the second dee when the
charges on the dees are suddenly reversed. This causes the particle to be
accelerated again by repulsion from the second dee and attraction back
to the first dee, all the while remaining on a circular path because of the
magnetic field. If all of this is done very quickly using radio-frequency
oscillation and the charges move in a vacuum without being absorbed,
one has created a "cyclotron." After many cycles of acceleration, the cyclotron produces a charged particle, or a bunch of charged particles, moving
at an extremely high speed or, equivalently, at an extremely high energy.
As the particles move faster and faster, they spiral outward toward the
outer rims of the dees. When they finally escape the dees, they move in a
straight line directed onto a target.

Lawrence and his assistant Stanley Livingston built their first working
cyclotron in 1931. It was small enough to fit in their hands. A year later,
using dees nine inches in diameter, they managed to accelerate positively
charged protons to energies in excess of 1 million electron volts (1 MeV).[3]
An electron volt is the equivalent energy attained by one electron accelerated across a potential difference, or voltage, of one volt. The higher the
energies of the accelerated particles, which were always positive, the more
easily they overcame the electric repulsion of the positively charged nucleus at the center of every atom, smashing into and even through it.

THE MIRACLE YEAR

Lawrence was among the first American physicists to turn to nuclear
physics. The breakthrough to quantum mechanics in 1925 had led by 1929
to quantum theories of magnetic materials, conductivity in metals, and
the chemical properties of elements and molecules arising from the orbiting
electrons in the outer structure of the atom. These successes spurred the
prospects of new commercial applications and encouraged much related

research in academic and industrial laboratories on solids, molecular and atomic spectra, chemical bonds, and electromagnetism. Among the over 250 research articles appearing in the volumes of *Physical Review* for 1930, the favorite topic of research by far was spectroscopy of all sorts, from fine and hyperfine structure, to spark and arc spectra, X-ray spectra, multiplet lines, Raman spectra, and nuclear and magnetic influences on molecular band spectra. Other popular topics were the electromagnetic properties of matter, radioactivity, electron behavior, and acoustics.[4]

One of the most intractable theoretical problems of the period arose from attempts to take the next step in the progress of quantum mechanics, the joining of quantum mechanics with relativity theory and electrons with electromagnetic fields in the formation of a new theory called relativistic quantum electrodynamics. Because the electrons in atoms move at very high speeds close to the speed of light, Einstein's special relativity theory must be taken into account. In 1928 British physicist Paul Dirac derived an equation, the Dirac equation, which provided a description of spinning electrons that satisfied both quantum mechanics and the theory of relativity. But when Heisenberg and Pauli attempted to join this equation with the quantum theory of electromagnetic fields, the theory exploded in mathematical infinities.

The infinities arose from the joining of the supposed infinitesimal size of the electron with the continuous, long-range electromagnetic field. Even a calculation of the energy of an isolated electron surrounded by its own electric field yielded an infinite result because the electron was assumed to have zero size, an infinitesimal point. If a minimum size for the electron is introduced, the infinity disappears, but then the theory no longer satisfies the requirements of relativity theory. The same dilemma arose for other charged particles and for other types of fields as they were introduced during the 1930s. Try as they might, theorists could not manage to avoid these infinities until over a decade later.

Despite the infinities, proposed quantum theories of nuclear phenomena could be studied and possible solutions to their problems tested in experiments in which the elementary particles approach each other at distances smaller than their tiny sizes—as, for example, when a high-speed proton smashes into another proton. But accelerators did not reach the necessary energies for these experiments until after World War II. Until then, physicists focused on two other areas of experimental research that involved events at extremely short distances in the search for clues that might lead to a future, problem-free theory.

The first area of research entailed collisions of high-speed cosmic rays with matter. Discovered early in the century, cosmic rays striking Earth's atmosphere from outer space were found to contain extremely high-speed particles and nuclei, apparently accelerated by cosmic events to energies far beyond any energies then attained in accelerators. With the outer structure of the atom now fairly well understood, the second area of research involved the tiny nucleus at the center of each atom, which now became a focus of attention.

The positive charge of the minuscule nucleus determined the number of negative electrons orbiting the nucleus of an electrically neutral atom, somewhat as the planets orbit the sun in our solar system. But there seemed to be more mass in the nucleus than the total mass of the protons. What could provide the remaining mass? Another question arose regarding the stability of the nucleus. Why didn't the nucleus fly apart under the electric repulsions among the positive protons crammed into the tiny space of the nucleus, about one ten-thousandth the diameter of the atom?

The answers to these questions began to emerge in 1932, a year sometimes called a "year of miracles." In addition to Lawrence's achievement of the 1 MeV cyclotron, James Chadwick at Cambridge University discovered the neutron. The neutron provided the beginnings of an answer to nuclear structure: neutrons and protons together constitute the mass of the nucleus, while only the protons account for its positive charge. (Neutrons possess no charge.) According to the quantum mechanics of nuclei developed by German physicist Werner Heisenberg (and independently by Russian physicist Dimitri Ivanenko), the protons and neutrons are all held together by a new force, much stronger than the electrical repulsion among the protons in the nucleus, hence its name, the "strong" or nuclear force. While it is very strong, it is also very short ranged, hence the tiny size of the nucleus. This attractive force also acts between all protons and neutrons. The nuclear force, Heisenberg suggested, is generated by an exchange of elementary particles, much as the atoms in some molecules are bound together by the sharing of identical electrons. But the exact origins of this force and the nature of the exchanged particles still remained a mystery.

Also during 1932, Harold Urey at Columbia University discovered the deuteron, a hydrogen nucleus (proton) with an extra neutron attached. When bound together with oxygen, two hydrogen atoms possessing deuterons as nuclei formed what was called "heavy water." Because protons and deuterons are both electrically charged, unlike the neutron, electric attraction and repulsion could be used to accelerate them to high speeds. They could then be smashed like bullets into target nuclei, and whatever debris emerged could be used as further clues to understanding nuclei, nuclear forces, and the particles themselves. This is exactly what John Cockcroft and Ernest Walton at Cambridge University did, initiating in 1932 the first disintegration of a nucleus by bombarding low-mass elements with high-speed protons.[5]

The deuteron also led to a valuable tool for nuclear studies, the Oppenheimer-Phillips process, named for Oppenheimer's first doctoral student at Berkeley, Melba Phillips. Phillips was born in Hazelton, Indiana, in 1907, graduated from high school at the age of 15, and received a bachelor's degree in mathematics from Oakland City College of Indiana in 1926. After receiving a master's degree in physics at Michigan's Battle Creek College in 1928, Phillips headed to Berkeley where she joined the small group of brave students studying theoretical physics under the demanding and often inscrutable J. Robert Oppenheimer. She completed her dissertation on theoretical analyses of the atomic spectra of the alkali metals and received her doctorate in 1933. At that time, Oppenheimer and his students were becoming more involved, as were other theorists working closely with experimentalists, in helping Lawrence and his collaborator Edwin McMillan understand their latest data. Lawrence and McMillan were bombarding nuclei with the newly discovered deuterons accelerated in their ever improving cyclotron. They found that the results agreed with the predictions of a theory developed by George Gamow. But when their cyclotron attained higher energies, the results disagreed with Gamow's theory. Phillips and Oppenheimer calculated that at energies above 2 MeV, the deuteron's proton and neutron will separate upon hitting the nucleus. The neutron will then be captured by the nucleus, but the proton will continue through it and emerge on the other side. Lawrence and McMillan confirmed this process, and in 1935 Oppenheimer and Phillips published the result, which became, wrote Hans Bethe, "an important tool in the study of nucleon energy levels and their properties."[6]

Phillips went on to an active, though difficult, career in physics education. In 1937 she joined the faculty of the Connecticut College for Women, and then transferred to Brooklyn College a year later. During the war, she did not join the Manhattan Project but instead was on leave from Brooklyn College to the University of Minnesota. In 1952 she refused to testify before a Senate committee searching for Communists during the McCarthy era and was fired from Brooklyn College (which later apologized for its action). After a period of unemployment, Edward Condon at Washington University in St. Louis, where Arthur Compton was now chancellor, appointed Phillips associate director of a teacher training institute at the university. In 1962 she joined the University of Chicago, where she developed a well-known series of courses on physical science for non-science majors. She was also well known for her popular textbook *Classical Electricity and Magnetism*, coauthored with Wolfgang Panofsky. In 1967 she became the first woman president of the American Association of Physics Teachers. Phillips died in 2004 in Petersburg, Indiana.[7]

Also during that miracle year of 1932, Carl Anderson, who had received his doctorate at Caltech in 1930, was researching under the direction of Robert Millikan the effects of cosmic rays careening into a thin slab of lead placed in a cloud chamber in his laboratory at Caltech. The cloud chamber could display the tracks of particles emerging from the collisions. In 1932 Anderson announced the discovery from this work of an entirely new particle that carried the mass of an electron but a positive charge, the opposite of the electron's negative charge. One surprising result of Dirac's relativistic Dirac equation of 1928 was the prediction that "antiparticles" should exist, that is, particles that carry the same mass but the opposite charge of their counterparts. With the help of theorist Oppenheimer, Anderson identified the new "positive" as the first antiparticle, the antiparticle to the electron, what is now called the positron.

About the same time as Anderson's discovery, Patrick Blackett and Giuseppe Occhialini in Manchester, England, discovered the existence of cosmic ray "showers"—a sudden burst of new particles created from the energy released when a single high-energy cosmic ray strikes a thin lead plate in a cloud chamber. The showering particles were identified as myriad electrons, positrons, and photons (light quanta), a phenomenon that, because of the problems with infinities, still seemed beyond the reach of existing quantum theories. Hampered by their limited theories, the theorists struggled to keep up with the flow of data from the experimenters' laboratories.

By the end of the decade, Anderson, Urey, Chadwick, and Lawrence had received Nobel Prizes for their work. Blackett received his prize in 1948. Isidor I. Rabi's work at Columbia in 1938 on a resonance technique for measuring the magnetic properties of nuclei by use of particle beams led to his Nobel Prize for physics in 1944. In 1937 Clinton J. Davisson at Bell Telephone Laboratories received the first Nobel Prize in physics awarded to an American industrial scientist for the discovery of the quantum wavelike diffraction of electrons by crystals.

THE CRISIS YEAR

The striking number of scientific discoveries, and the prospects of more to come so long as research funding continued, contrasted sharply with the simultaneous collapse of the economy in the United States and worldwide. It reached crisis proportions in that same miracle year. The Wall Street crash of 1929 wiped out the endowments of most private universities. But many others that could rely on private foundations and state funds managed to limp along until 1932, when industry, banking, and agriculture joined the collapse. The national unemployment rate jumped to 25 percent. (For comparison, during the recession beginning in 2008, it hovered officially at 9.6 percent.) Overall, federal budget cuts to science amounted to 12.5 percent, but the budget for the National Bureau of Standards, which employed numerous physicists in support of industry, dropped by 50 percent, leading to the layoff of half its staff. Industrial research, which had reached a funding peak in 1931, hit rock bottom in 1933. General Electric also laid off half of its researchers, and AT&T laid off 40 percent.[8]

As states tightened their belts, research at state universities felt the pinch. Perhaps typical of state universities, the budget for the Berkeley physics department dropped 18 percent during the period from 1929 to 1932, then another 21 percent the following year. The panicked department laid off a quarter of its teaching assistants and slashed the salaries of those who remained by nearly a third. Professors' salaries suffered an

8 percent cut.[9] Because of the low job prospects through 1935, graduate students who could do so hung on longer, and on little or no pay. Postdocs, some with families to support, faced destitution. Many left the field, but others found support through the creative funding efforts and the connections of their elders.

The world economic crisis hit Germany especially hard. The rise of Hitler and his National Socialist (Nazi) Party to power in January 1933 was soon followed by a wave of dismissals of Jewish teachers, professors, and other civil servants, including many who were part-Jewish or married to Jewish spouses. Most of the dismissals occurred during the crisis years of 1933 to 1935 in the United States. According to one count, 278 physicists and 162 mathematicians were summarily fired from academic jobs in Germany. Of these, 192 had found their way to the United States by 1939.[10] Among them were many of the best and brightest scientists of the day, including the past or future Nobel Prize winners Maria Goeppert-Mayer, James Franck (whose future wife Herta Sponer, a physicist, had also emigrated to the United States), Hans Bethe, Felix Bloch, Otto Stern, Eugene P. Wigner, and, of course, Albert Einstein.

Recognizing the direction of events in Germany during the 1920s, Einstein, then a professor in Berlin, had accepted Millikan's invitation to an annual part-time visiting professorship at Caltech beginning in 1930. He was in the United States in a half-time position at the Institute for Advanced Study in Princeton, New Jersey, when Hitler came to power. As the dismissal policy came into effect, he was one of the few physicists anywhere to speak out forcefully against the new regime, and he angrily resigned all of his distinguished positions in Germany. He soon settled full time at the Institute for Advanced Study. The Institute was founded in 1930 and directed until 1939 by medical educator Abraham Flexner, with the financial support of the department-store magnate Louis Bamberger and his sister Caroline Bamberger Fuld. In addition to Einstein, Kurt Gödel, John von Neumann, and other leading émigrés found refuge at the Institute, making it one of the world's premier research facilities in mathematics and theoretical physics during the decades ahead.[11]

Physicists in the United States and abroad attempted to aid their displaced colleagues through donations and the establishment of a refugee network. With Bohr, Pauli, and Einstein among its most active players, the network facilitated emigration from Germany by helping to place physicists in temporary university positions abroad, with funding provided by the Rockefeller Foundation and the donations of wealthy individuals such as Oppenheimer. Unfortunately, the nation's immigration quotas were still in place, as were the quotas on the admission of Jewish students to universities. Most universities were not very eager to welcome many refugees to their faculties. In addition to already high academic unemployment, anti-Semitism was still prominent in higher education.[12] "It is disgusting to notice that good students can not find employment in America only because they happen to be Jewish," Samuel Goudsmit, a Dutch émigré, complained to Harrison Randall in 1938. While most universities did not want more than a few token Jewish faculty members, age discrimination was also a factor. Einstein wrote to Pauli in the same year: "It is extraordinarily difficult to find positions. No faculty will hire a man over 50 unless he has accomplished something extraordinary—and of course a Jew not at all."[13]

Despite discrimination and the economic hurdles, many refugee physicists, including a number of women scientists, did find academic and industrial positions in the United States. With their enthusiasm and European training, they helped push the already extant peaks of American physics even higher. Because many of the refugee physicists were theorists (owing in part to laboratory discrimination in Europe), American theoretical physics reached new heights at a time when the forefront of research in quantum theory concerned phenomena associated with the high-energy collisions of cosmic rays with matter and the smashing of nuclei by charged particles in accelerators. After the arrival of the refugees, there was now no doubt that, as *Newsweek* declared in 1936, "The United States leads the world in physics."[14]

Yet just three years earlier the collapse of the American economy had unleashed a public backlash against science, especially against physics, that nearly brought research to a halt. The partnership that Millikan, Hale, Jewett, and other managerial physicists had carefully forged with industrial corporations and private foundations during the 1920s had been so successful that President Herbert Hoover, a former commerce secretary, had run for office in 1928 on a platform promoting an even stronger alliance of government with science managers and business leaders as the foundation of the nation's future prosperity. More science meant more commercial applications, which in turn meant more jobs. But in the wake of the Wall Street crash and the rise in unemployment that ensued, an angry public turned the formula around: they blamed physics, together with industry, for helping to create the crisis. Overwhelmed by new consumer products, the public believed that progress in physics had led to the overproduction of technical goods and to the invention of too many labor-saving technologies that had put people out of work. Critics called for a "moratorium on research." Science, they argued, had become too dependent on industries generating products and profits with little concern for their economic and social consequences, and too dependent on private donors who imposed their own agendas on research.[15] The advance of pure science, it seemed, had outpaced the economic, social, and even ethical needs of society.

Reformers demanded that scientists modify their allegiance to private wealth and enterprise through renewed commitments to solving the nation's economic and social problems. Warren Weaver, the director of natural sciences at the Rockefeller Foundation, agreed. In 1933 he declared that the physical sciences no longer needed the foundation's financial support. Instead, he announced, funding of NRC fellowships for physics postdocs would be slashed and the funds channeled into the less developed medical and social sciences that had more direct impact on human needs. Projects in the physical sciences would be funded only if they contributed to solving problems in these fields.[16]

The public backlash, the poor economy, and Weaver's change of policy did not deter the ever enthusiastic Ernest Lawrence at Berkeley. Nor did it deter corporations from funding pure accelerator research in the hope of reaping lucrative future patents from the "atom smashers." In 1932, as Berkeley's budget collapsed, the university president provided Lawrence a large grant of $5,170, along with an old campus shack to house his new cyclotron invention. Lawrence named it the Radiation Laboratory or "Rad Lab" for short. Although Rockefeller funding to the Rad Lab declined to just $500 for NRC fellows during the years 1931 to 1933, the nonprofit Research Corporation, founded to handle research patent rights and to dispense the profits as research grants, had been funding cyclotron construction from the start. (Lawrence patented the devices himself but assigned the patents to the corporation.) As Rockefeller funds declined, the corporation picked up the slack with a grant of $7,500, to which was added $4,500 from other private sources.

Despite the depression, the Research Corporation also began placing its bets across the board, making a series of large grants for accelerator construction to Cornell, Chicago, Columbia, and Purdue. By 1934 the Rad Lab was smashing atoms with 1.5 MeV protons and 3.0 MeV deuterons. In the same year, Cornell built its first cyclotron. A year later, cyclotrons at the other three recipients of Research Corporation funds came on line, as did four others funded by other philanthropies and industrial grants.[17] While research in other areas of physics lagged for lack of funds, there was clearly no depression in accelerator physics.

ADMINISTRATORS RESPOND

The political winds suddenly shifted at the federal level when Franklin Delano Roosevelt defeated Republican candidate Herbert Hoover in the presidential election of 1932. As unemployment soared in 1933, the new president saw little chance that accelerators, or any other pure-physics research, could immediately generate the large numbers of new jobs now so desperately needed. As part of his New Deal to get the economy back on its feet as quickly as possible, Roosevelt slashed federal funding to physics research in both government and academe and transferred the funds into relief programs. Yet, because of its increased support to the much needed agricultural and biological sciences, the government still remained the largest contributor overall to the nation's research programs, providing roughly half the nation's total expenditure for science.[18]

With broad areas of academic and industrial physics now under economic, social, and political pressure, adept science administrators once again rose to the occasion by adapting their ideology of pure science to the new environment. Their task was to find a way to modify the partnership of science with private industry and philanthropy so as to encourage the well-funded, socially oriented government programs to support academic research. Yet at the same time they still wanted to avoid, as before, the "political trickery" of government intrusion into the sanctity of "pure science." The chief architects of this undertaking were the president of

MIT, Karl T. Compton, and the ever active physicist and CEO of Caltech, Robert A. Millikan. Both men had been closely associated with Hoover's policy of promoting science as a spur to private enterprise. Their efforts presented a challenge to Roosevelt's New Deal progressivism by raising a conflict between the direct social engagement of science for national benefit demanded by Roosevelt and the indirect, "trickle-down" approach of pure science leading eventually to new technologies, and hence to new jobs and more prosperity.[19] Millikan told a national radio audience in 1934: "Leave the human spirit free for the development of science and education, and no bounds can be set to the possible fullness of the life of the average citizen of the United States in the coming century."[20] Their failed attempt to manage this conflict helped to prepare the way for a new partnership with government that later emerged in the midst of the coming world war.

Like Lawrence and Millikan, Karl T. Compton was the product of a Protestant family and a midwestern university. He was born in 1897 in Wooster, Ohio, the eldest of four brothers and one sister, one of whom was the Nobel laureate Arthur H. Compton. Karl attended Wooster University (now Wooster College) where his father was a philosophy professor. He graduated in physics in 1908 and headed a year later to Princeton University. There he received the doctorate in physics in 1912 for experimental work under Owen Richardson on the electron theory of metals and the quantum photoelectric effect. In 1915 Compton became a professor at Princeton and a consultant to General Electric Laboratories. He encountered Millikan during World War I while working with the Signal Corps in Washington, D.C., as part of the Research Information Service. After the war, Compton continued his research at Princeton on electrons and spectroscopy, including the passage of photoelectrons through metals and gases. According to a biographer, his gift for teaching became legendary at Princeton, "marked not alone by the clarity of presentation and a contagious enthusiasm, but also by his manifest concern for the well-being and progress of each student."[21] He was chair of the Physics Section of the National Academy of Sciences during the first half of the Hoover administration. Following his appointment as president of MIT in 1930, a post that he held until 1948, Compton became an advocate for stronger independent scientific research at MIT as the basis for a stronger engineering education and for the eventual benefits of industrial progress.

It was in pursuit of these latter objectives that Compton accepted his appointment by Roosevelt in 1933 to the chairmanship of the new Scientific Advisory Board (SAB), an arm of the National Academy of Sciences. Its purpose was to act as a mediator between government sponsorship and the private control of science. Like the NRC during World War I, the SAB functioned through the appointment of committees of politically "disinterested" elite scientists, none of whom—to critics' dismay—were involved in government research or policymaking. Among its fifteen members in 1935 were, in addition of course to Robert A. Millikan, representatives of academic and industrial research. They included Frank B. Jewett, now president of Bell Telephone Laboratories; Charles F. Kettering, president of the General Motors Research Corporation; the astronomer and former president of the University of California William Wallace Campbell, now president of the National Academy of Sciences; Isaiah Bowman, chairman of the NRC; and Abraham Flexner, director of the Rockefeller Institute of Medical Research and of the Institute for Advanced Study.[22]

In September 1933 Compton and the SAB submitted a "recovery plan" for science to the U.S. Interior Department for the huge sum of $16 million to be spent over six years in support of basic research. But in order to avoid what Jewett called "political interference" and "a large measure of bureaucratic control," the SAB demanded that the distribution of the funds to private researchers at elite institutions be left to "disinterested, non-partisan minds."[23] Harold Ickes, the Secretary of the Interior, vetoed the plan. He could not see how the funding of a few elite universities and professors would provide enough jobs for the nation's many unemployed scientists, to say nothing of the general population. In the end, the Rockefeller Foundation funded the board through its new Division for Social Sciences.

During the next two years, as the SAB responded to requests for advice and recommendations from government agencies on science-related policy issues, it faced increasing accusations, as did the earlier NRC, that it was merely a bunch of independent academic and industrial scientists attempting to influence the creation of federal policy. It also came into increasing conflict with a newly formed competitor, the National Research Board chaired by Interior Secretary Ickes. The National Research Board's inclusion of social scientists and its emphasis on projects of immediate public benefit made Compton's SAB seem even more out of touch and even more superfluous as an advisory body. When its mandate expired without renewal in 1935, the SAB quietly went out of business.[24]

LEARNING LESSONS

The failure of the SAB caused considerable political fallout. It left the National Research Board, now called the National Resources Board, with an even freer hand to pursue not just research support but active research planning and direction. In 1935 its social scientists, including the new historians and sociologists of science, began an investigation of the nature of scientific research and its relationship with government benefactors. The result was a widely influential report published in 1937 titled "Research— A National Resource." Science was seen as an essential factor underlying the future economic and cultural power of the nation. Government should thus take significant responsibility for fostering and guiding this resource for the good of the nation.[25] President Roosevelt went even further. With the lessons of chemical warfare and the criticism of science in service of unbridled industrial profits apparently still in mind, he declared during his inaugural address for a second term in 1937 that only through federal intervention could the United States hope "to create those moral controls over the services of science which are necessary to make science a useful servant instead of a ruthless master of mankind."[26]

Karl T. Compton and his colleagues also learned their lessons from the failure of the SAB. Even as they still resisted the influence of government agendas on any aspect of research, Compton was now calling in 1935 for a "bilateral program" of the nation's scientists and government leaders "for putting science to work for the national welfare."[27] But he and his colleagues also realized that such a program would work to their favor only if scientists, especially physical scientists, cooperated with the government from a position of strength, a strength derived from the tangible benefits their work could provide for the government's immediate purposes. Four years later, the outbreak of war in Europe provided the scientists just such a basis of strength for obtaining federal cooperation on their own terms. As American entry into the war with Germany and its allies seemed more likely, Compton and his colleagues among the scientific leadership began arguing that physicists could and should serve the nation, that the nation should prepare to enter the coming war, and that government should find new federal funding for research, "a national resource." If before the war physicists had benefited the nation through new consumer products, after the outbreak of war they could benefit it with new military "products," including, as we shall see, the most ominous product of all to emerge from the nuclear laboratories, the unleashing of nuclear fission.[28]

THE LATE THIRTIES

By 1935 the worst of the economic crisis had abated. The economy had stabilized. Although it had not yet fully recovered, industrial research funding was back to precrisis levels. Overall, university research funding approached its previous high by 1936. Under Roosevelt, the federal government became not only the largest benefactor of American research but also its most influential policymaker. Programs like the Works Projects Administration (WPA) created jobs of all types, including those for numerous scientists and graduate researchers. In 1939 the National Advisory Committee for Aeronautics, still active since World War I, sponsored research contracts for twelve special investigations at ten research universities.[29]

The physics department at Purdue University was perhaps typical. With funding from government and private foundations, and the insights of a steady stream of visiting fellows and temporary and permanent new faculty from abroad, the department launched a range of new research programs in 1936. These included precision acoustics, optical spectra, electronics, theoretical physics, and nuclear research with the construction of both a Van de Graaff accelerator and a cyclotron with 37-inch dees producing 8.3 MeV deuterons. But the overall budget for research did not reach its previous high until the end of the decade. At Cornell and MIT, work in similar fields related to industrial applications flourished, together with the construction of a cyclotron and a Van de Graaff machine.[30] An accelerator was by then a must-have piece of equipment for any serious physics program in the United States.

Faculty salaries in the physics department at the University of California at Berkeley were also back to previous levels by 1936, but the overall department budget did not recover until 1938. The department did not hire any new faculty until 1939, which caused its own crisis. While some students had waited out the job shortage by staying in school, the economic revival caused a sudden surge of new students. By 1938–1939 the department was swamped with eighty graduate students, of whom thirty-five were newcomers. Previously, the total number of students had hovered at half that number.[31] Fortunately for the students, Berkeley also hosted Lawrence's ever expanding accelerator laboratory. The Rad Lab was by then a university department, helping to educate and employ large numbers of physics graduate students and postdocs. By 1939 the Rad Lab, perched in the manner of a medieval castle on an expansive hill overlooking the town and campus, employed over 200 workers of all ranks and skills.

While physics departments experienced the ups and downs of the Depression, the structure of the profession underwent profound changes. Progress in quantum mechanics and experimentation were now pushing the forefront of physics further into the realm of technical quantum equations, matched by equally sophisticated experimental apparatus. Theorists and experimentalists drew even closer of necessity. At first, the theorists primarily assisted the less mathematically adept experimentalists in understanding the new theories and their own data. But as the data and the theories grew ever more complicated, the two disciplines joined together in an equal partnership of mutual benefit. This occurred in Europe as well as in the United States. Yet, as Sylvan S. Schweber has shown, the more cooperative and open atmosphere of the American physics department enabled an especially close and fruitful collaboration among theorists and experimentalists, and this was no more apparent than in "the collaborative, cooperative, and collective nature of the organization of the cyclotron laboratories."[32]

Although considerable funding for research in all sciences flowed through government programs during the late 1930s, seemingly impractical fields like nonaccelerator nuclear physics and quantum theory tended to receive less federal funding. But these were the fields that were advancing most rapidly after 1935. Private foundations once again filled the gap, but less so the corporations whose patience for new patents had apparently run out. The national resource that was science now seemed to administrators of the Rockefeller Foundation sufficient cause to reinvest in abstract physics. The "titanic tool of science," as the foundation president called it, could be used to strengthen American democracy and to serve humanity by counteracting the rise of antiscientific dictatorships abroad.[33]

Lawrence and others following his lead managed to extract large sums from the Rockefeller Foundation and other sources, including the National Advisory Cancer Council (NACC), which provided federal funds, by successfully arguing the benefits of bigger accelerators for the advancement of medical research. These funds were in addition to state subsidies.[34] The accelerators could be used, they argued, to create radioactive isotopes for radiation therapy and to produce particle beams to target the tumors directly. So much money began flowing into the Rad Lab and other sites that accelerator construction in the United States became an even bigger boom enterprise during the late thirties. Other medical research and more practically oriented fields such as electronics and solids once again languished both in funding and in status.

By 1940 American physicists were running or building twenty-four cyclotrons as well as about ten other accelerators. The former included the biggest machines yet built, the 60-inch cyclotrons under construction at Berkeley's Rad Lab and at the Carnegie Institution in Washington, D.C.[35] The vast majority of these were built and run by physicists trained in Lawrence's laboratory, which was itself setting new records of cost as well as size. During the academic year 1937–1938, the Rad Lab's support amounted to $87,600, which, in addition to state funding ($45,900), was derived from the first "medical" grant of $15,000 from the Rockefeller Foundation and lesser sums from the WPA, the Research Corporation, and the Macy Foundation.[36] That year, the serious-looking maker of the big machines captured the cover of *Time* magazine.[37]

A NEW PARTICLE

Even with this spectacular success, Ernest Lawrence was eager for more. No sooner was his Nobel Prize in hand and his 60-inch machine spewing out 16 MeV particles than he laid plans for an even bigger cyclotron, a colossal machine measuring 184 inches across. The purpose of the colossus was to explore the latest discoveries in nuclear and particle physics.

Many of those discoveries were stimulated by a new theory of nuclear forces put forth in 1935 by the Japanese physicist Hideki Yukawa.

Yukawa attempted to solve the mystery of exactly what held the protons and neutrons together within the nucleus by proposing that the strong attractive force arises between and among protons and neutrons through the exchange of a new particle possessing a charge equal to that of a single electron charge but a predicted mass of some 200 times the mass of an electron. The heavier mass was required to account for the enormous strength of the nuclear force. Yukawa also suggested, vaguely, that "it may also have some bearing on the shower produced by cosmic rays."[38]

Yukawa's theory remained unknown in the West until 1937 when Oppenheimer and his assistant Robert Serber in Berkeley brought it to Western attention. In that year, their colleagues at Caltech, cosmic ray experimenters Carl Anderson and Seth Neddermeyer (and, independently, J. C. Street at Harvard and his collaborator E. C. Stephenson), discovered yet another new particle among cosmic ray events in their detectors. The new particle possessed the same charge as an electron (whether positive or negative) but a mass about 140 times greater than that of the electron. They called it the "heavy electron." After learning of Yukawa's work, quantum theorists in the U.S. and abroad attempted to link the heavy electron with Yukawa's heavy exchange particle, what they eventually called the "meson." In so doing, they linked experimental high-energy physics with theoretical nuclear physics, theory with experiment. But in order for experimentalists to produce mesons through collisions, accelerators required energies in excess of 100 MeV. Only a huge machine such as the one proposed by Lawrence could possibly achieve those high energies.

As war engulfed Europe in the spring of 1940, the Rockefeller Foundation, still convinced of the civilizing influence of fundamental physics, awarded Lawrence's Radiation Laboratory the huge sum of just over $1 million for the construction of his giant machine. The remainder of its cost came from the university. Just two years earlier, the Rad Lab's financial support had amounted to $87,600.[39] The award and the planned machine marked the culmination of the decisive turn in American physics to "big science," a reliance upon big machines, built and run on big budgets, with huge staffs encompassing everyone from machinists and technicians to Nobel Prize theorists and experimentalists, all working toward the promise of big payoffs in comprehending the most fundamental aspects of nature. In hindsight, the turn to collaborative big science, as epitomized in the big-machine physics of the Rad Lab, also presaged and prepared the way for not only the even bigger machine physics of the postwar era, but also for one of the biggest big-science projects of all time, a project that emerged just over a year later: the Manhattan Project to build the atomic bomb.

4
The Physicists' War

On August 2, 1939, Albert Einstein signed a letter to President Franklin D. Roosevelt in which he stated:

> Sir: Some recent work by E. Fermi and L. Szilard, which has been communicated to me in manuscript, leads me to expect that the element uranium may be turned into a new and important source of energy in the immediate future. Certain aspects of the situation seem to call for watchfulness and, if necessary, quick action on the part of the Administration.

Quick action might be necessary, Einstein continued, because

> this new phenomenon would also lead to the construction of bombs, and it is conceivable—though much less certain—that extremely powerful bombs of a new type may thus be constructed.[1]

Eight months earlier, researchers in Berlin had announced to the world the discovery of nuclear fission, the splitting of the uranium nucleus accompanied by the release of a large amount of energy. (As with the word *nucleus, fission* derived from cell biology.) By the summer of 1939 it was known in theory that atomic fission might result in a nuclear explosive of unsurpassed power. But this was possible only if a sufficient amount of the very rare form of uranium, the isotope uranium-235 (U-235), could be assembled into a "critical mass," the minimum mass needed to sustain a nuclear explosion. The splitting of a U-235 nucleus was set off by the absorption of a neutron, which is not repelled by the positive nucleus. It was discovered that when a U-235 nucleus fissions into two smaller nuclei, it releases not only energy but also more neutrons, two to three on average. These neutrons could then go on to split more U-235 nuclei, each of which producing more neutrons. A chain reaction occurs. If the ball of uranium is large enough and dense enough that the reaction continues for many steps, so much energy is released so quickly that an explosion of enormous energy occurs.

In 1939 no one yet knew for certain if such a chain reaction would indeed occur. Nor did they know how much U-235 was needed to attain a critical mass, nor the best process for extracting the extremely rare isotope U-235 from natural uranium ore, nor how exactly to set off the explosive chain reaction. But, Einstein informed the president, they did know that Germany was busily acquiring uranium in Europe and that German scientists, who had discovered fission, were hard at work in Berlin on exploiting their discovery. Still, the likelihood that anyone could build a fission bomb in the near future seemed very remote.

Less than a month after Einstein sent his letter to the president, Hitler unleashed German panzer divisions onto Poland, igniting the war in Europe. In October 1939 the president, inspired by Einstein's letter, established a small advisory committee at the National Bureau of Standards to study the prospect of utilizing nuclear fission. Not until the Japanese attacked Pearl Harbor on December 7, 1941, bringing the United States to the war against Japan, Germany, and their allies in what became World War II, did Roosevelt finally authorize a crash program to build the bomb. But by then, thanks to the familiar efforts of able science administrators, a large portion of the physics community was already mobilized and ready to join in support of the war effort in many areas, including the building of the bomb. If World War I had been the chemists' war, World War II would be the physicists' war.

ADMINISTRATORS TAKE COMMAND

Among the able administrators who leapt into action were the familiar players: Karl T. Compton; Robert A. Millikan; Isaiah Bowman, now president of Johns Hopkins University; Frank B. Jewett, now president of the National Academy of Sciences and of Bell Laboratories and vice-president of AT&T; and Karl Compton's brother Arthur, the Nobel Prize physicist at Chicago. But the lead role fell to one of the most able science administrators of the period, indeed of the century, Vannevar Bush, the president of the Carnegie Institution in Washington, D.C., a research institute and administrative arm of the Carnegie Endowment. James Bryant Conant, an organic chemist and president of Harvard University, served as Bush's right-hand man.[2]

Born in 1890 to a middle-class family in Everett, Massachusetts, Vannevar Bush was the grandson of two sea captains. His father was a minister in the Universalist Church. (There is no known relation with the presidential Bush family.) The younger Bush was, writes a biographer, "pragmatic, yet had the imagination and sensitivity of a poet, and was steadily optimistic."[3] Bush was educated in engineering at Tufts College and received an engineering doctorate in 1916 in a joint program with MIT and Harvard. During World War I, he had worked in the antisubmarine research laboratory at New London, Connecticut, sponsored by the National Research Council (NRC). Returning to MIT after the war as an electrical engineering professor, Bush and his students invented a calculation device for solving sixth-order differential equations that is considered a forerunner of the modern computer. Bush rose to dean of engineering and vice president of MIT during the early 1930s under MIT president Karl T. Compton.

As an adherent of Hoover's ideal of the public-spirited corporate technocrat, Bush, like Compton, Jewett, and other leaders, opposed government meddling in scientific and business matters through the New Deal. But this did not hinder his good relations with Roosevelt's White House. In 1936 he was appointed head of the NRC Division of Engineering and Industrial Research and in 1939 to the chair of the National Advisory Committee for Aeronautics (NACA), the federally funded committee for military aviation research.[4]

While Bush, Compton, Bowman, and others sought beneficial new relationships for science with the federal government following the demise of the Scientific Advisory Board, Millikan offered the military the full services of the NRC in a collaboration reminiscent of that during World War I. The military refused. It had its own laboratories and the NACA to fund university and corporate research on military-related matters. Having experienced first-hand the benefits of this type of collaboration during World War I, Vannevar Bush, supported by Millikan, took up the cause. He remained undeterred by military reticence as he sought new ways to integrate science and engineering into military research as both a boost to science and a support for the nation. The outbreak of war in Europe in 1939, together with the discovery of nuclear fission, suddenly gave these men the ammunition they needed. In addition, most physicists, incensed by the persecution of scientists and the suppression of free thought by foreign dictators, were willing to prepare for military research, even if it required major compromises with the humanistic, progressive ideals still held by most scientists and the general public.[5]

Bush, Conant, and colleagues were worried that the United States was once again falling behind its European scientific competitors, especially Germany, in scientific advances and in the development of new technological weapons. The lessons of gas warfare in the last war were still fresh in their minds. Bush had already established the small Advisory Committee on Uranium with the president's approval after the receipt of Einstein's letter. Although the United States was still officially neutral, Conant pushed for war preparations in a meeting with Bush, Jewett, and others in Washington, D.C., in May 1940. The nation, Conant argued, was mired in dangerous "isolationism," and its leaders were unaware of the benefits that science and technology could bring in time of war. Just as his predecessors before America's entry into World War I, Bush went straight to the president the next day to obtain his support to begin mobilizing American science and technology for the nation's probable entry into the war. On June 14, 1940, Roosevelt approved the formation of the National Defense Research Council (NDRC), chaired by Bush, tasked with preparing civilian science for military research.

As with prior committees, NDRC members included primarily civilian scientists: Harvard president Conant, MIT president Karl Compton,

Caltech dean of science Richard Tolman, and the president of Bell Labs and the National Academy of Sciences Frank B. Jewett. To these were added Conway P. Coe, the commissioner of patents, and an army general and a navy rear admiral. A year later, Roosevelt, again at the request of Bush and Conant, absorbed the NDRC into the new and larger Office of Scientific Research and Development (OSRD) directed by Bush for the coordination of the nation's research in support of military applications. Conant took command of the NDRC within the OSRD organization.[6]

Although other federal committees emerged to challenge the OSRD, Bush successfully defended his organization as the one bearing prime responsibility for the research and development of new military applications.[7] Rather than putting the scientists in uniform, as occurred during the previous war, Bush borrowed from the models of the NACA and the NRC and, again despite the earlier appeals to pure science, readily enlisted civilian university and industrial laboratories to the cause through federal contracts to undertake specific military research projects.

Most of the laboratories funded or created through the NDRC or OSRD were located at the same elite universities that had received the bulk of federal research funds and fellowships during the previous decades. In fact, according to one assessment, the OSRD spent 90 percent of its funds for academic contracts at just eight institutions.[8] They, and leading corporate laboratories, were now equipped and staffed at the highest levels possible. Among the recipients of the new federal largesse flowing from the mobilization program were MIT's Radiation Laboratory for the development of radar, Caltech for the development of solid-fuel rockets, Johns Hopkins for the proximity fuse, the University of Chicago's Metallurgical Laboratory for nuclear reactor design and construction, and Lawrence's Radiation Laboratory for the study and separation of fissionable isotopes. Purdue University received a smaller contract to use its cyclotron for isotope separation as well as a subcontract in support of radar development.

Among the industrial laboratories, Western Electric, a subsidiary of AT&T, received the largest corporate funding, followed by DuPont, RCA, and General Electric. By the time of the Japanese attack on Pearl Harbor, it is estimated that 1,700 physicists were already working on war-related research. Lawrence's Rad Lab alone employed 142 physicists, of whom nearly all were engaged in fission-related research.[9]

Bush and Conant were well on their way to success in their effort to create a new model for the relationship of science, especially physics, with the military and political power centers of American society. It was a relationship that, despite the avoidance of political influence, entailed the integration of research with military needs. Once again, the aversion to political influence did not extend to the military, mainly because most scientists regarded the military as nonpolitical, even though its influence on research might be even greater, while the ideology of humanistic, "disinterested" pure science was not needed, or wanted, in time of war. Looking to the future, the stage was already set for the postwar era.

ESTABLISHING THE CHAIN OF COMMAND

Not until shortly before Pearl Harbor did nuclear fission become a top priority for the civilian scientist-administrators. Controlled fission in a reactor was likely to succeed soon, and theoretical research had pointed to the possibility of a bomb, but Conant, Millikan, Lawrence, and other leaders doubted that a bomb would prove technically feasible in this war. That view began to change in October 1941 when the United States received a secret British report on fission prepared primarily by two German refugees working in Great Britain, Otto Frisch and Rudolf Peierls. Code-named the Maud Report, it concluded that indeed "a uranium bomb is practicable and likely to lead to decisive results in the war." The report recommended "that this work be continued at the highest priority and on the increasing scale necessary to obtain the weapon in the shortest possible time."[10]

Bush once again went straight to the top. On October 9, 1941, he presented the British report during a meeting with President Roosevelt and Vice President Henry Wallace. The president immediately approved exploratory research on building the bomb under the auspices of the newly established OSRD. But not until a month after Pearl Harbor did Roosevelt sign a letter drafted by Bush approving work in preparation for building the bomb. And not until March 11, 1942, after Bush had submitted a progress report on the awesome power of the bomb and a possible race with Germany for it, did Roosevelt approve a crash program to build the new weapon. In a handwritten note sent to Bush, the president simply wrote,

"The whole thing should be *pushed* not only in regard to development, but also with due regard to time. This is very much of the essence."[11] Yet not until April 1943 was the Manhattan Project finally under way. Bush and the American administrators were still overcoming their skepticism about the feasibility of a nuclear weapon.

After Roosevelt's approval of bomb exploration, in January 1942 Bush reorganized the OSRD to bring scientists into closer collaboration with the two military branches at that time, the army and the navy. Within days of Pearl Harbor, he had already placed the Advisory Committee on Uranium, now called Section S-1, under the oversight of James B. Conant. Under Bush and Conant, the work of Section S-1 split among three research teams. Arthur Compton and the Chicago Metallurgical Laboratory, the "Met Lab," took responsibility for the fundamental physics, which included the theoretical research group on uranium fission under the direction of Oppenheimer at Berkeley. Lawrence was assigned research on the electromagnetic separation of fission isotopes using the huge magnets of his new Berkeley cyclotron, while Harold Urey at Columbia investigated gaseous diffusion as a method for extracting the needed rare uranium isotope for an atom bomb.[12] Thus, by early 1942 the nation's entire fission research effort rested squarely under Section S-1, which was under Conant and his vice director, Richard Tolman, of the NDRC, which was under the OSRD, which was headed by Vannevar Bush, who stood directly beneath the President.

The military-style chain of command was intentional. After all, the nation was at war. But, while working to bring scientists into partnership with the military, Bush also arranged for them to occupy a dependent and subordinate position within that partnership regarding research policy and responsibility for their work. Only military and political leaders and science administrators, mainly himself, were accorded any voice at all in the overall direction and use of the research. Bush reported in a letter to Conant that during his meeting with the president and vice president on October 9, 1941, Bush had asked the president to issue an order "to hold considerations of policy on this matter within the group consisting of those present this morning, plus Secretary [of War] Stimson, [Army Chief of Staff] General Marshall, and yourself [Conant]."[13] When Arthur Compton raised a policy question soon thereafter, Bush responded in an authoritative tone: "the problem which is placed before your committee is the technical problem and not the problem of what should or should not be the governmental policy in this program."[14]

It was the beginning of another important turning point in the relationship of physics to the political, military, and even corporate power centers of society. Under Bush and Conant, the pure-science ideology of the scientist as the responsible keeper of moral culture and an equal partner with other important groups in society was not only dropped, but replaced with a much more limited and subordinate conception of the scientist. It was a conception reflected in the OSRD's system of contract research.[15] Instead of regarding the scientist as an elite, disinterested researcher of physical processes standing above practical research, Bush and the OSRD now viewed the project scientist as little more than a technician of nature fulfilling a contract, a worker relieved of any responsibility for the direction of the research or its use.

THE OTHER RAD LAB

The Manhattan Project functioned as a subunit of the civilian-run OSRD and its S-1 committee. Under the directorship of Vannevar Bush, the OSRD pursued the usual policy of funneling federal contracts mainly to the most prestigious university and corporate laboratories. Because much of the military-related research concerned electronics, rocketry, nuclear fission, and related topics regarded as engineering applications of fundamental physics principles, the vast majority of the contracts went for physics research. According to one estimate, in 1942 the OSRD spent four times more on physics than it did on chemistry.[16] Aside from atom bomb development, one of the biggest recipients of OSRD funding was the electronics laboratory at MIT. It was deliberately named the Radiation Laboratory, or "Rad Lab," after its Berkeley counterpart in order to confuse the enemy and outsiders. One of the biggest successes to come out of the Rad Lab was the development of microwave "radar," an acronym for *ra*dio *d*etection *a*nd *r*anging.[17]

Radar was already a reality before the outbreak of war, but its meter-length radio waves were prone to interference and unable to detect low-flying aircraft. The development of a radar system using 10-centimeter-length microwaves seemed the best alternative for detecting, identifying, and navigating aircraft and ships. By early 1940 the Wall Street tycoon and amateur physicist Alfred Loomis was at work in his private laboratory on microwave radar under contract with the NDRC. Despite the heroic efforts of Loomis, Lawrence, and others, the work was not going very well when, in the fall of 1940, the British exported to the United States a new invention, the "cavity magnetron." The device, a resonator, promised to produce microwaves in the 10-centimeter range with sufficient power to generate an effective radar beam. The goal suddenly seemed within reach just as the German Luftwaffe began its assault on London in an effort to bomb Britain into submission.

Bush's NDRC, the predecessor of the OSRD, awarded nearly half a million dollars to MIT for a project employing roughly fifty physicists to develop microwave radar. Karl Compton's connection with MIT and the institute's long-standing work with government and industry were strong factors in its favor.[18] Lawrence recommended Lee DuBridge, chairman of the physics department at Rochester University, to head the new Radiation Laboratory. Because cyclotron builders were familiar with the uses of resonant electromagnetic waves for the acceleration of particles in cyclotrons, ten of the first members of the MIT Rad Lab were cyclotron workers, including several of the top "cyclotroneers" from the Berkeley Rad Lab.[19]

By spring 1941 the physicists of the MIT Rad Lab had a prototype microwave radar device ready for testing. Unfortunately, it failed to meet army aircraft specifications. But this early prototype was suitable for another use by the navy: the aircraft detection of German submarines when they surfaced for air and battery recharging. Its successors proved more successful in meeting the army's needs.

After the United States entered the war in December 1941, the military demands on the MIT Rad Lab for new electronic hardware increased dramatically, along with its budget. Within a year, the laboratory had a staff of 2,000 and a budget of $1.15 million. By the end of the war, the staff had reached nearly 4,000 members, of whom 500 were physicists, and it occupied 15 acres of floor space in and around Cambridge, Massachusetts, and maintained offices in several countries abroad. Its total wartime funding of $1.5 billion was second only to that of the Manhattan Project, which was about $2.2 billion.[20]

To keep up with the demand, the MIT Rad Lab established the Research Construction Corporation outside Boston for the production of prototypes. Among the products assembled, many in collaboration with British scientists and engineers, were advanced microwave radar for detection and navigation; radar jamming and evasion devices; and a system for long-range navigation. This new system consisted of a network of crossed beacons in the sky to enable planes and ships to determine their locations to an accuracy of 1 percent.

Freed from university obligations, the physicists of the MIT Rad Lab eagerly embraced the excitement of cutting-edge research, stimulating teamwork, and the sense that they were performing a useful task for the defense of their country. It was a formula that few could resist, even after the war had ended. DuBridge later quipped that the atomic bomb ended the war, but radar won it.[21] The MIT Rad Lab devices and many of the scientists who invented them were involved in practically every major Allied military operation of the war. During the D-day invasion of Europe in June 1944, an advisory group of physicists successfully jammed German coastal radar and provided radar beacons for the paratroopers' drop zones.

BUILDING THE BOMB

Ernest Lawrence was in the midst of building his monster cyclotron when Bush and the S-1 committee asked him to begin work on the separation of the rare U-235 isotope from natural uranium. Because the extremely rare isotope was chemically identical to the other, more plentiful uranium isotopes, the usual chemical means of separation would not work. Instead, Lawrence used the cyclotron as what is known today as a mass spectrometer, a device often employed for identifying chemicals and forensic evidence.

In the retooled device, the electrically charged uranium nuclei moved through the magnetic field of the cyclotron magnets and experienced a force perpendicular to their direction of motion. This resulted in the bending of the paths of the nuclei into a curve. But the amount of curvature differed according to the mass of the nuclei. Because of this, the various isotopes of uranium, including U-235, were directed onto slightly different curved paths according to their slightly different masses, thus allowing experimenters to separate U-235 from the other uranium isotopes.[22]

Lawrence, the experimentalist, had already worked closely with Oppenheimer, the theorist, on problems of nuclear structure. After nearly a year of collaboration on isotope separation at Berkeley, Lawrence had Oppenheimer invited to a secret conference on fast-neutron fission to be held in October 1941 at the General Electric Research Laboratory in Schenectady, New York.[23] Arthur Compton, chair of the conference, was so impressed with Oppenheimer's command of the theory of bomb design that he appointed him to lead the fast-neutron research unit at Berkeley. In May 1942 Compton promoted Oppenheimer to director of the nation's entire theoretical research effort on nuclear fission. The task was to combine theoretical calculations with the scant available experimental data on uranium metal and the fission process in order to estimate the required critical mass, the energy yield, and other information required to construct the bomb. Oppenheimer organized a summer research session in the Berkeley physics department to explore the prospects. In addition to several of Oppenheimer's assistants, a number of the nation's top theorists participated, the majority of whom, like Hans Bethe and Edward Teller, had immigrated from Europe.

By that time, researchers at the nearby Berkeley Rad Lab had already discovered two new "transuranium" elements, elements beyond uranium (element 92) on the periodic table. They were later called neptunium (element 93) and plutonium (element 94) after the planets beyond Uranus. Both of these new elements are unstable, very fissionable, and easily produced as by-products of a working reactor. But only plutonium was stable enough to be used as a substitute for uranium to power an atomic bomb.[24]

Oppenheimer reported to Arthur Compton at the end of the summer that, in theory at least, a nuclear reactor and a uranium bomb were feasible and that once a reactor is running it could be used to produce the easily obtained fuel for a plutonium bomb. But the realization of either bomb still "would require a major scientific and technical effort."[25] He also reported on his committee's finding of the prospect of an even more powerful weapon: a fusion or hydrogen bomb, called the "Super." Teller was eager to continue exploring the prospects of the Super, but Oppenheimer recommended, much to Teller's displeasure, that work on the Super should be put aside until after the war.

Bush and Conant eagerly welcomed Oppenheimer's report as the final scientific justification required, in addition to Roosevelt's approval, to launch a crash program to build the bomb. On September 17, 1942, the

Army Corps of Engineers, which had been assigned responsibility for the bomb project, promoted Colonel Leslie R. Groves, the "can-do" builder of the Pentagon, to the rank of Brigadier General and commander of the Manhattan Engineer District (named for the location of its early office) to build the atomic bomb. The District included the newly formed Manhattan Project, the central laboratory charged with designing and building the bomb from components produced at other locations of the so-called District.

The general traveled to Berkeley to consult with Lawrence and Oppenheimer on the task ahead. To everyone's surprise, at the end of October 1942, Groves appointed the unlikely Oppenheimer as the scientific director of the Manhattan Project. The appointee had little experimental ability, no administrative experience, and a questionable leftist political past. But Groves saw in him a man who could quickly grasp the entire range of a problem, command the respect of other physicists, and display as much dedication as Groves to making this project a success. Even more important, Oppenheimer's political vulnerability meant that, under Groves's protection, he was unlikely to challenge Groves's authority as commander.[26]

Upon Oppenheimer's recommendation, Groves selected a remote site near Santa Fe, New Mexico, then occupied by the Los Alamos Boys School, as the location of the Manhattan Project, the central laboratory for the design and assembly of the atomic bombs. The other components of the effort under Groves's command included Compton's Met Lab at the University of Chicago, where the world's first nuclear reactor went critical under Enrico Fermi's direction in December 1942. They also included the Clinton Laboratories at Oak Ridge, Tennessee, which contained huge industrial facilities for the separation of fissionable U-235 and the production of heavy water (used as an alternative to graphite for reactor construction); and the facility near Hanford, Washington, where the DuPont chemical company, the inventor of nylon for parachutes and stockings, designed, built, and ran plutonium-producing "breeder reactors" over the objections of physicist Eugene P. Wigner, who was working with the Chicago team. (Wigner's own doubtful design would not have worked because of impurities.)

In 1943, at the insistence of the scientists, the University of California, rather than a military agency, was selected to act as the institutional contractor for the work at Los Alamos.[27] Lawrence, Oppenheimer, and many of the workers of the Berkeley Rad Lab brought the lessons of big science with them as they transferred to Los Alamos. Despite some objections, the university's role as the sole joint contract manager of both of the nation's nuclear weapons development sites—Los Alamos and, later, the Lawrence Livermore Laboratory—has continued almost to the present. It now shares that responsibility.

By the end of the war, the Manhattan Engineer District employed over 200,000 people, making it the world's largest and—at $2.2 billion—most expensive research and development effort until the advent of the Apollo Space Program, which landed a man on the moon during the 1960s. People from everywhere in the country and all walks of life contributed to the effort, from pipe fitters, welders, and Nobel Prize physicists, to the machine operators at Oak Ridge known as the "Tennessee Girls," and the local Native Americans who served as maids and babysitters for the Los Alamos scientists.[28] In June 1943 British engineers began to arrive at Los Alamos to aid in the effort as well.

Everyone involved was relatively young: most were in their twenties. Isolated in the New Mexico wilds and with many newlyweds among the workers, there was a veritable baby boom across the laboratory. As at the MIT Rad Lab, all of this contributed to a sense of excitement, a common bond with others sharing the difficulties and hardships of family life in the ramshackle houses, and a dedication to the common goal of winning the war. There was a universal feeling that this was a very special time in their lives. Even today, the veterans of Los Alamos often look back upon those days much as a later generation would look back upon Woodstock.

When Los Alamos finally got under way in April 1943, Oppenheimer divided the laboratory and its work into four divisions: theoretical physics, experimental research, chemistry and metallurgy, and ordnance. A fifth division was established for Teller with the task of planning for postwar projects, mainly the Super. It was not only Teller's favorite research topic, but Oppenheimer had apparently also sought to console Teller for having selected Hans Bethe instead to head the theoretical physics division.

One of the greatest remaining difficulties involved the triggering of the bombs. The uranium bomb could be set off by joining together two subcritical pieces of U-235 into a critical mass. But the joining had to occur extremely rapidly, or stray neutrons in the air would initiate the chain reaction even before the critical mass was assembled, causing the bomb to fizzle. The British suggested using a cannon inside the bomb to shoot one hemisphere of uranium toward another at high speed. But this would not do for plutonium, which is so fissionable that it would begin to explode, then fizzle, even at the speed of an artillery shell.[29]

As fissionable uranium and plutonium began arriving at Los Alamos from Oak Ridge and Hanford, testing revealed that the cannon design would indeed work for uranium. Design and building of the uranium bomb was soon under way. For plutonium, however, a very sophisticated rapid-implosion design was essential. Teller, John von Neumann, and Seth Neddermeyer, apparently borrowing from Tolman, who had gotten the idea from the implosion deaths of stars, hit upon an arrangement involving a critical mass of plutonium shaped into a spherical shell at very low density, surrounded by an outer shell of conventional high explosive. Upon ignition, the high explosive would implode the plutonium extremely rapidly into a tiny ball of dense critical mass, setting off a nuclear explosion. But the implosive compression of plutonium had to occur under a precisely spherical shock wave, or else the resulting lopsided critical mass would yield only a minor eruption. If a spherical design could be made to work, plutonium would be more suitable than uranium for the production of an arsenal of nuclear weapons, owing to the relatively easy acquisition of plutonium from the breeder reactors now pumping out the highly fissile material at the Hanford site.[30]

With the implosion design and the arrival of the British engineers, the pace quickened at Los Alamos. Ever determined to achieve success, General Groves was eager to deploy the new weapon as soon as possible. As early as March 1944, three months before the D-day invasion of German-occupied France, he mobilized the Army Air Force to begin preparations for dropping the atomic bombs on Germany and Japan. In September, Groves ordered production schedules for the delivery of uranium and plutonium; in January 1945 Colonel Paul W. Tibbets, selected to pilot *Enola Gay*, the B-29 bomber that would drop the uranium bomb on Hiroshima, was already preparing his flight crew for its fateful task.

DROPPING THE BOMBS

Because the solution to the implosion problem still seemed doubtful, the
Los Alamos scientists petitioned the navy captain in charge of the ordnance division for a precombat test of the plutonium bomb. None was
needed for the uranium bomb, which they were certain would succeed.
The captain hit the roof. He complained to Groves that by requesting a
test the scientists displayed an interest only in doing scientific research,
and that this test would delay the dropping of the bomb. Oppenheimer
attempted unsuccessfully to defend the scientists. Only after a visit to the
laboratory by James Conant and a letter from him approving the test did
the scientists at the bottom of the command chain receive their wish.[31]
The successful test of the plutonium bomb, the first nuclear detonation
in history, occurred on July 16, 1945, at the so-called Trinity test site in
the New Mexico desert near Alamagordo, about 250 miles south of Los
Alamos. The next day, a similar plutonium bomb was on its way to the
Pacific to join the uranium bomb for use on Japan.

As plans hastened for the Trinity test, the momentum of the work and
the pressure to complete it as soon as possible built to such an extent that
any doubts at Los Alamos about the project's ultimate goals were overcome
by the sheer excitement of the science and the rapid progress of the work.[32]

Still, the possibility remained that Nazi Germany would be the first to
achieve a nuclear weapon, with consequences too horrible to imagine. But
by the end of 1944, as American and British forces began smashing their
way into France and Germany after D-day, it was evident that the Germans did not have the bomb and that Germany would be defeated before
the Allied bomb was ready. Joseph Rotblat, a Polish refugee engineer on
the British team, quietly resigned from the project. There was no longer
any need for the bomb, he felt: "The whole purpose of my being at Los
Alamos ceased to be, and I asked for permission to leave and return to
Britain."[33] Permission was granted. Although others had avoided joining
the Manhattan Project, and some now questioned the purpose of the project, Rotblat was the only one to resign. He later received the Nobel Peace
Prize for his postwar work toward nuclear control and disarmament.

As the building of the bombs proceeded at Los Alamos, General Groves
appointed a joint military and civilian Targeting Committee chaired by

Oppenheimer. Its task was to select the Japanese cities to be targeted and to determine the procedure for dropping the bombs, including the optimum height of detonation in order to achieve the maximum possible devastation. Physicists were then dispatched to the Pacific to arm the bombs for detonation in flight over Hiroshima and Nagasaki. In the dropping of the bombs, physicists were not just the providers of new technical weapons but willing partners in their use.

Matters took a different turn at the Chicago Met Lab after it had completed its assigned tasks. Following the German surrender in early May 1945, objections to use of the bomb on Japan grew louder. They found expression in two important documents that emerged from Chicago. One was a petition circulated by Hungarian refugee Leo Szilard and submitted to the newly installed President Harry S. Truman on July 17, 1945, the day after Trinity. (Roosevelt had died in office on April 12, 1945.) The petition called upon the president to consider "the moral responsibilities which are involved" and to offer the Japanese an opportunity to surrender rather than being subjected to attack without warning. Otherwise, the petition continued, a surprise attack would set a dangerous precedent for future nuclear warfare. The president, however, had already left Washington for a meeting of Allied leaders in Potsdam, Germany.[34]

The second document, the so-called Franck Report, emerged from the Chicago laboratory's Committee on Political and Social Problems, established at the Met Lab by its director, Arthur Compton. The chair of the committee, Nobel physicist James Franck, a refugee from Nazi Germany, submitted the report to the secretary of war on June 11, 1945.[35] Like the Szilard Petition, the Franck Report opposed a surprise nuclear attack on Japan. It also called for a public demonstration to the Japanese of the power of the bomb; and it warned of a postwar nuclear arms race if the United States used the bomb.

But, equally important, the Franck Report may be seen as an attempt by the scientists to regain control of their work. "In the past," they declared, "scientists could disclaim direct responsibility for the use to which mankind had put their disinterested discoveries." But this is no longer possible with nuclear weapons, they wrote, "which are fraught with infinitely greater dangers" than past inventions. Of course, the report declared, "the scientists on this Project do not presume to speak authoritatively on problems of national and international policy." But, they continued, scientists are among the small number of people who do have knowledge of the technical aspects of these weapons and of the "grave danger for the safety of this country as well as for the future of all the other nations, of which the rest of mankind is unaware. We therefore feel it our duty to urge that the political problems, arising from the mastering of nuclear power, be recognized . . . and that appropriate steps be taken for their study and the preparation of necessary decisions."[36]

In the way of bureaucracies, the secretary of war passed the Franck Report to his Interim Committee on nuclear issues, headed by Bush and Conant, who passed it to their committee's Scientific Advisory Panel, consisting of Oppenheimer as chair, Ernest Lawrence, Enrico Fermi, and Arthur Compton. The advisory panel found no feasible alternative to a surprise nuclear attack on Japan. A demonstration bomb might prove to be a dud, they reasoned, thus causing the opposite effect. If warned of an impending attack, the Japanese might put prisoners of war in the target area. Instead, the committee supported an argument put forth by Compton and others at the time who emphasized "the opportunity of saving American lives by immediate military use."[37] It is estimated that a D-day style Allied invasion of the Japanese homeland would have cost upward of a million lives, including Japanese as well American and those of the other Allies.[38]

Regarding policy matters, the advisory panel went even farther. In an important statement for the postwar era, Oppenheimer, writing for the panel, renounced the Franck Report's insistence on a measure of responsibility by the scientists for the use of their work. He reaffirmed Bush's vision of scientists as contractors providing technical results with no role in decision making concerning their use. "With regard to these general aspects of the use of atomic energy, it is clear that we, as scientific men, have no proprietary rights. It is true that we are among the few citizens who have had occasion to give thoughtful consideration to these problems during the past few years. We have, however, no claim to special competence in solving the political, social, and military problems which are presented by the advent of atomic power."[39]

On July 26, ten days after the successful Trinity test, the leaders of the United States, Great Britain, and China issued an ultimatum to the Japanese demanding unconditional surrender. "The alternative for Japan is

prompt and utter destruction," the ultimatum declared—without elabo-
ration.[40] The Japanese rejected it. On August 6, 1945, the uranium bomb
obliterated 68 percent of the Japanese port city of Hiroshima, along with
many of its residents. Three days later, the plutonium bomb devastated
Nagasaki with another great loss of life. It is estimated that in the range of
200,000 people died in the two blasts and another 100,000 died later of
injuries, burns, and radiation poisoning.

The Japanese government received another shock on August 9 when the
Soviet Union declared war on Japan and invaded Manchuria, then occupied by Japanese forces. On August 15, Japan sued for peace. On September
2, Japanese representatives surrendered unconditionally. The physicists'
war was over.

5
Taming the Endless Frontier

The stunning successes of the Manhattan Project, the MIT Radiation Laboratory, and the many other research and development efforts during the war convinced the nation's leaders of the crucial importance of fundamental discoveries achieved through what was now called basic research. The close collaboration of the military with scientists and engineers working in the highly technical disciplines of nuclear physics, electromagnetic theory, and electronics had produced the war's "winning weapons."

As victory approached, President Roosevelt asked his top science administrator Vannevar Bush to reconnoiter the contours of the postwar relationship between science and the federal government. In his well-known and widely influential report submitted to President Truman in 1945 titled *Science: The Endless Frontier*, Bush, still director of the Office of Scientific Research and Development (OSRD), argued not only that the close partnership must continue, but that a devastated Europe could no longer provide the new fundamental knowledge on which the successful wartime technologies had largely rested. The federal government must now take an active role in funding and promoting basic research in civilian laboratories, and those laboratories must be willing to accept federal funding. In the pure science tradition he defined the basic research to be promoted as that performed "without thought of practical ends." Despite its federal funding, the sponsored research would entail the curiosity-driven exploration for new knowledge of nature on the endless frontier of science.[1] This knowledge would eventually find its way into new beneficial applications. By not exploring that endless frontier, he argued, the nation would place itself at a competitive disadvantage, militarily and economically.

Federal officials already expected that new discoveries in basic physics would continue to yield benefits for "national security"—not only by enabling the development of new weaponry but also by enhancing the nation's scientific prestige in the increasing competition for power and recognition with the Soviet Union. Industry, too, expected potential postwar contributions of unfettered pure physics to the development of new products for the booming consumer society. But government still needed encouragement toward active promotion. Before the war, business and philanthropy had funded basic research in the big-science accelerator laboratories in the expectation of new patents and medical cures. The federal government, Bush argued, must now take on this role. Even as physicists remained suspicious of government political influence, they understood the necessity of federal funding for large-scale projects whose costs soon exceeded the means of universities and private donors and even most corporations. Physicists needed the government as much as the government needed physicists. But the partnership had to be redefined as the nation entered the postwar era.

THE NATIONAL SCIENCE FOUNDATION

One avenue to the redefined partnership ran through Bush's report to the president. The Bush report is best known for its main proposal, the establishment of a new federal agency to replace the OSRD and to institutionalize Bush's vision of the postwar science-government partnership. It led to what became today's National Science Foundation (NSF). Drawing upon elements of prewar pure science and his own wartime success, Bush presented four main arguments for the new foundation: the essential need for basic research as the foundation of future power and prosperity; the vital role that government funding must play in fostering civilian research; the need for a civilian-directed national science policy, including a military research policy; and, finally, the freedom of inquiry for basic research and researchers from government policy meddling, even as universities accepted huge sums in support of the federal policy agenda.

In order to achieve these aims, Bush proposed a civilian "National Research Foundation." As did the wartime OSRD, the new foundation would act as an intermediary. It would funnel federal research funds directly to universities and other nonprofit laboratories "that," he wrote, "should by contract and otherwise support long-range research on military matters."[2] If Bush had his way, physicists and other basic researchers would be freed as before from political influence and obligations regarding their work, but, also as during the war, they would be subject instead to the demands and obligations of long-range research on "military matters." In return they would achieve secure federal funding; civilian administrators would exercise influence equal to federal authorities over the supported research; and, in the long run, the nation would reap the competitive benefits.

As early as 1945, Senator Harley M. Kilgore, a New Deal Democrat, submitted a bill for the creation of what was now called the National Science Foundation to replace the OSRD—but not exactly along the lines Bush had in mind. The new NSF would fund research and education in all fields of science and medicine, including civilian military research, but it would also include the social sciences. Under the authority of an administrator appointed by the president, Kilgore's NSF would also establish and coordinate a national research policy, but it would direct grants as well to applied research for the social good. It would also vest all patent rights from funded research in the federal government rather than in private hands, and it would spread its funds evenly across the country and across research institutions in order to raise the quality of all.

The Kilgore approach was straight out of the New Deal. In opposition to it, Senator Warren Magnuson, also inspired by Bush, submitted an alternative to the Kilgore bill in 1945. The Magnuson bill replaced the president's appointee as the overseer of the foundation by an autonomous Science Board of civilian scientists in order to insulate the foundation from political influence. (But Congress would still hold the purse strings.) In addition to the contract system, the NSF would operate through the project-grant system invented by German scientists during the 1920s, and for much the same reason: to insulate science from the German democracy of the period. In this system, researchers submitted project proposals for competitive evaluation through independent peer review and approval by the independent Science Board. It was a process that would prove as highly successful for the Americans as it had for the Germans. In addition, the Magnuson bill, with the concurrence of the Science Board, stripped from Bush's plan the funding of military-related civilian research. It also excluded funding for the social sciences, which Bush regarded as politically motivated. Despite President Truman's concerns about Bush's own political motives, a compromise NSF bill embodying much of Bush's vision, except for civilian military research, finally passed Congress, and Truman signed it into law in 1950.[3]

In its operation, the NSF exhibited the familiar strategy of pushing the existing peaks higher—the channeling of funds to large numbers of selected individuals at a small number of top universities, located mainly on the east and west coasts. Because those scientists were already among the elite, they naturally submitted the most competitive proposals for funding. In 1954–1955, for instance, 62 percent of NSF grants went to doctoral and postdoctoral researchers at just eleven institutions, all housing large research groups led by a few big-name scientists. This emphasis on supporting an elite meritocracy of researchers extended throughout the federal funding scheme for science in the United States, including physics. But because the NSF did not handle military research, Defense Department agencies quickly emerged as by far the nation's largest source of federal funds for research and development thereafter. According to NSF statistics, in fiscal year 1951, the Defense Department provided nearly 70 percent of federal research and development (R&D) funds, about $1.3 billion, while the entire budget for the NSF amounted to only about $150,000 (see Table 2 in the Appendix). During the academic year 1952–1953, there were ninety physics PhD granting institutions in the United States, but 72 percent of all federal funds for nonclassified academic research in physics went to just seventeen institutions enrolling 65 percent of the nation's physics graduate students.[4]

Still, as in the 1920s and 1930s, pushing the peaks higher once again achieved its purpose. It brought huge success to American physics in discoveries and growth during the 1950s and 1960s, and it maintained the nation at the forefront of world research in what many now regarded as the "American century," even if only a fraction of American physicists, and even fewer female physicists, could participate in the new discoveries.

THE MILITARY TAKES COMMAND

Having learned the lesson of the atom bomb, whose origins lay in seemingly arcane nuclear research, most military leaders required no convincing about the potential military value of "pure" science, even if they could not immediately foresee any useful applications. Funds began to flow into basic research almost as soon as the war ended. General Groves was the first to leap into action. In the fall of 1945 he provided $175,000 in leftover Manhattan Project funds for a new "synchrotron" accelerator at Berkeley. Lawrence's big 184-inch machine was capable of reaching the world-record energy of 100 million electron volts (MeV), but it hit a wall erected by relativity theory. As accelerated particles increase in speed, or kinetic energy, they also increase in mass, as required by Einstein's theory of special relativity. Because of the increasing masses of the accelerated particles, it is difficult to keep them on the cyclotron's circular track. Drawing upon an idea put forth by Australian physicist Marcus Oliphant, Lawrence's right-hand man Edwin McMillan solved the problem by altering the strength of the magnetic field and the frequency of the accelerating electric field in synchronization with each other and with the increasing masses of the particles. Thanks to Groves's generosity, by 1946 Berkeley's new "synchrotron" was up and running and producing particle energies exceeding 200 MeV. This was more than enough energy needed to produce new particles—and new discoveries—out of the energies of accelerated particles smashing into targets.[5]

The Groves grant served as both a reward for wartime service and a down payment on any future discoveries of potential military value. Not to let the army gain an advantage, in 1946 the navy opened the Office of Naval Research (ONR) in Arlington, Virginia. It began doling out millions in contract funds for basic and applied research to nearly every serious researcher who asked, and with few strings attached.[6] The idea that independent "pure research" would inevitably lead to practical applications required little argumentation. For the navy there was no question that curiosity-driven research was as potentially beneficial to the military as was applied research and development. The ONR became the primary supporter of the nation's academic research laboratories, and American scientists became the best funded of any in the postwar industrialized world. The ONR support to physicists in their home laboratories became so ubiquitous that nearly 80 percent of the papers presented during a meeting of the American Physical Society in 1948 acknowledged ONR support.[7] So much money flowed into research from military and nuclear-funding agencies that some physicists began to worry about the public perception of physics. Lee DuBridge, Millikan's successor as president of Caltech, told a congressional committee, "There is a wide-spread feeling in the country that the *only* purpose of science is to develop weapons of war and that science can be kept on a wartime footing . . . The chief goal of science is *not* to develop weapons, but to understand nature."[8]

Massive military funding helped drive the rapid expansion of American science, promoted the growth of computer technology, fostered new hybrid disciplines such as geophysics, and supported important foundational studies in fields such as physical meteorology and global warming. But few scientists apparently bothered to consider the potential effects of military funding, even without visible strings attached.[9] Nevertheless, such funding did come at a price for the scientists and their science. By accepting federal defense funds, the scientific community could not easily object to military plans that it might find objectionable, including the later program to build the hydrogen bomb. Nor could the defense-supported scientific community easily object to the heavy-handed treatment of its members by McCarthy era inquisitions, imposed loyalty oaths, and the laboratory secrecy required by the national-security state.

Most importantly, however, historian Paul Forman has argued that the generous military funding of science caused "a qualitative change in its purposes and character." Impressed by the wartime successes of radar, rockets, and the atomic bomb, military funders tended to emphasize technological superiority over fundamental new scientific insights. The effects, Forman argued, could be observed, for example, in the study of quantum electronics, which brought us the laser: "we may say that support by military agencies and consultation on military problems had effectively rotated the orientation of academic physics toward techniques and applications . . . Physicists had lost control of their discipline."[10]

Other historians, most notably Daniel Kevles, have disputed the "distortionist" argument that military funding "seduced American physicists from, so to speak, a 'true basic physics.'" Instead, Kevles argues, such funding exerted a positive influence, not only by promoting the rapid expansion of physics, but also by helping to integrate American physics into the national-security system as both a research and an advisory enterprise, where it enjoyed greater influence in promoting its interests.[11] Others have perceived the possibility of a middle position arising out of a "'grey area' in the distortionist debate": scientists were able to maintain a measure of independence even as their institutions engaged in classified research or were heavily funded by the military.[12]

As tensions with the Soviet Union increased after the war, the nation continued its weapons programs while maintaining science and engineering on a permanent war footing. Appropriate institutions to manage these activities now became essential. While the ONR provided one channel of military funding and the NSF another for nonmilitary funding to universities, a new organization was needed to replace the Manhattan District and to oversee all the nation's nuclear research. In October 1945, the Truman administration submitted to Congress the May-Johnson Bill for the establishment of an Atomic Energy Commission (AEC). Named for the two senators who sponsored the bill, it was largely the product of army administrators, including General Groves. The bill was also supported by a small group of physics leaders—Lawrence, Fermi, Oppenheimer, and Arthur Compton. Its provisions placed control of all nuclear research in a part-time commission of military officers appointed by the president. It emphasized secrecy and the military control of research; it called for the continued development of nuclear weapons over economic uses of nuclear energy as the primary goal of American policy; and it incorporated only few patent protections against the industrial monopolization of nuclear power and related applications by a few leading corporations.[13]

Most physicists greeted the May-Johnson Bill with shock and anger. Nuclear weapons, they argued, should be controlled by a civilian agency, nuclear power should also benefit civilian energy needs, and civilian research should not be completely under military control. The bill unleashed a storm of protest, galvanizing an already growing scientists' movement for the international control of nuclear weapons. Moreover, it forced an irreparable break between the scientists and their leaders in Washington that began to spell the end of the long tradition of a few elite scientist-administrators exercising authority over the affairs of the entire physics community.

Protesting scientists descended upon Washington, and organizations including the Federation of Atomic Scientists, the Association of Los Alamos Scientists, and the influential publication *Bulletin of the Atomic Scientists* began mobilizing public opinion against the bill. Harold Urey told Congress that the May-Johnson Bill "would create a potential dictator of science." Leo Szilard said the bill seemed aimed at only one purpose: "to make atomic bombs and blast hell out of Russia before Russia blasts hell out of us."[14]

Surprised at the physicists' response, Truman began to entertain alternatives. In December 1945, after working with the scientists, Senator Brien McMahon, with Truman's support, submitted a bill for an AEC composed instead of a civilian director and five full-time civilian commissioners.[15] It would include advisory committees of civilian scientists and engineers, funding for nonclassified basic research in addition to nuclear weapons research (the first commercial reactors did not appear until 1951), and provisions for any patents resulting from federal funding to be held by the federal government rather than by private individuals and corporations. The bill passed in June 1946 and was signed into law at the end of the year. The AEC remained in place until 1974, when it was split into the Nuclear Regulatory Commission (NRC) and the Energy Research and Development Administration. In 1977 the latter became today's cabinet-level Department of Energy, responsible, among other things, for maintaining the nation's nuclear arsenal. The NRC, which still oversees reactors and radiation, has remained an independent federal agency.

Truman appointed David Lilienthal, the director of the Tennessee Valley Authority, the hydroelectric power agency, to head the new AEC. One of Lilienthal's first acts was to appoint the commission's top advisory panel of civilian experts on nuclear weapons and reactors, the General Advisory Committee. It reflected the new working partnership among academic physicists, industrial engineers, and government officials in matters of nuclear policy, civilian and military. Chaired by J. Robert Oppenheimer, the other members of the General Advisory Committee in 1949 included Fermi, Conant, Rabi, DuBridge, Berkeley chemist Glenn Seaborg, industrialist Hartley Rowe, Cyril Stanley of the University of Chicago, and Hood Worthington, a DuPont engineer.[16]

Although the AEC was still primarily a nuclear weapons agency subject to presidential and military oversight, the scientists had achieved their goal of establishing civilian input regarding nuclear weapons policy. Between the ONR and the AEC, even more money began flowing into research with even fewer strings attached. Scientists were thinking again about big science. Because accelerators were still considered a branch of nuclear physics, the AEC inherited from Groves and the army oversight of accelerator physics. The leading physicists on the General Advisory Committee, most of whom were veterans of the Manhattan Project, convinced the AEC to begin pouring funds into the construction of expensive new accelerators. The accelerators might have seemed to offer little immediate practical military or commercial value, but Lawrence had used his 1940 accelerator to separate the first batch of fissionable uranium isotope—as fission physics had demonstrated, who knew what might possibly emerge, even from this highly abstract branch of physics?

With generous funding from the AEC, in 1947 Isidor I. Rabi, Norman Ramsey, and a consortium of nine universities founded an East Coast accelerator laboratory at Brookhaven on Long Island, New York. By 1948 the AEC was funding the construction of, and fostering competition between, even bigger synchrotrons at Brookhaven and Berkeley designed to reach 3 billion and 6 billion electron volts, respectively. In 1952 Edward Teller and Ernest Lawrence founded the Lawrence Radiation Laboratory at Livermore, California, not far from its parent, Lawrence's Rad Lab in Berkeley. With Herbert York as director, the new laboratory was intended to serve as competition for the Los Alamos laboratory. Thanks to AEC dollars, by 1953 the United States had two nuclear weapons laboratories, thirty-five accelerators, and eleven research reactors in operation, the most and biggest of any nation. The United States now led the world in civilian and military nuclear research and in the prestigious field of what was now called high-energy physics.[17]

EXPLOITING ATOMS FOR PEACE

Confident of the newly proclaimed American Century, the United States did not hesitate to use its enormous postwar scientific and technological power, together with its economic and military might, to promote its foreign policy interests abroad. At the end of World War II, most of Europe lay in ruins, economies were in collapse, and populations struggled without adequate food and shelter. One of the four postwar occupation powers in Germany, the United States appropriated any German patents, equipment, and scientists that it found useful for its domestic science and industry, as did the other powers. But as the four powers began to split into Cold War opponents, it was clear that an economic and political power vacuum in Western Germany and Europe might easily invite Soviet attempts at control or even occupation. American Marshall Plan money began flowing into the reconstruction of Western Europe, not only for humanitarian purposes but more so for creating a joint American-European alliance against Soviet domination. Naturally, the prevailing argument in the United States that the funding of basic research was essential to future economic and military growth applied also to European nations. While Vannevar Bush and others helped direct Marshall Plan funds into basic research in Europe, Isidor Rabi, who served on an advisory committee to the U.S. occupation authorities, worked toward the reconstruction and revival of German science as an aid to the revival of Western Europe as a whole.

As the U.S. representative to the United Nations Educational, Scientific, and Cultural Organization (UNESCO), Rabi presented a resolution to a meeting of the UNESCO general assembly in June 1950 calling for the establishment of "regional research centers and laboratories." Because West Germany was still prohibited from engaging in nuclear fission research of any type, including reactors, Rabi suggested the creation of a consortium of Western European nations—much like the Brookhaven Laboratory's consortium of universities—for the support of a single European high-energy accelerator laboratory that could, with initial U.S. assistance, compete on an international level with Soviet and American accelerators.[18] After overcoming initial suspicions about American motives, the Europeans eventually accepted, and in 1952 eleven nations created the

Conseil Européen pour la Recherche Nucléaire, or CERN, to plan and develop an accelerator facility to be constructed on the French-Swiss border near Geneva, Switzerland. In 1954 the Conseil changed its official name to Organisation Européenne pour la Recherche Nucléaire, or European Organization for Nuclear Research, but it kept the original acronym, CERN. (Future references here will retain current usage: CERN, the European Organization for Nuclear Research.)

The efforts to revive European science, and to create CERN as the centerpiece of that revival, helped to put Western Europe back on its feet. However, in the Cold War era, writes historian John Krige, these efforts also served "to promote a U.S. scientific and foreign policy agenda in Western Europe": to integrate Western Europe into an Atlantic alliance under the control and protection of the American nuclear umbrella. Even though Rabi and colleagues may not have had quite those aims in mind, the building or rebuilding of European laboratories aided the American agenda, not only by reviving a strong western Europe, but also by enabling American scientists to benefit from any interesting European research and by helping to overcome domestic hostility to U.S. support of foreign research as part of this agenda.[19]

The utilization of the U.S. lead in science and technology as a foreign policy instrument became clearer in December 1953 when President Eisenhower announced to the General Assembly of the United Nations a new American initiative toward the achievement of world peace, what he called "Atoms for Peace." Following the end of the Korean War and the death of Stalin in 1953, Eisenhower and his State Department settled on the Atoms for Peace initiative as a means to showcase American superiority and to promote United States nuclear policy for Europe while at the same time diffusing European concerns about that policy and pressuring domestic industries to invest in nuclear reactor technology.[20] In his address to the United Nations, Eisenhower called upon the Soviet Union and other nuclear nations to work together with the United States to reduce nuclear tensions and to redirect nuclear energy toward "the peaceful pursuits of mankind." Toward that end, he called, among other things, for the U.N. to establish a new Atomic Energy Agency (now the International Atomic Energy Agency, or IAEA) that would gather donations from nuclear nations of radioactive isotopes and enriched reactor-grade uranium to be distributed to non-nuclear nations. Reactors installed under the auspices of the agency would "provide abundant electrical energy in the power starved areas of the world," while the isotopes, distributed on a grander scale than was the case in any earlier U.S. program, would benefit medicine and agriculture.[21]

Eisenhower's Atoms for Peace program received nearly universal acclamation. Foreign nations welcomed it, both for its peaceful uses of atomic energy and for the economic and scientific benefits it afforded. The American public welcomed it as well, and much good did come of the international cooperation and collaboration. But from the longer historical view, Atoms for Peace again primarily served the United States' Cold War agenda for Europe. At that time, the United States was shifting from a reliance on conventional forces to defend Europe against a potential Soviet invasion to a less costly nuclear defense that required the nuclearization of the North Atlantic Treaty Organization (NATO) nations under American oversight.

But not all nations were eager to comply with the American nuclear agenda. Britain was already nuclear, and France was developing its own nuclear arsenal, and both had their own foreign policy plans. Prominent West German scientists successfully mobilized German public opinion against an independent German nuclear weapons program. Germany, the likely battleground if the Soviet Union did invade Western Europe, has remained militarily non-nuclear ever since.

The building of peaceful reactors in Europe helped to divert foreign public concern and reluctance during the controversial transition to reliance on nuclear deterrence. At the same time, the delivery of isotopes for research and peaceful applications helped to ensure American access to foreign research for monitoring and utilization, while the arrangement of international scientific meetings provided rare opportunities for assessing the capabilities and activities of Soviet scientists and those from other nations for intelligence purposes.[22]

Isidor Rabi, then chairman of the AEC's General Advisory Committee, initiated and helped organize the most famous of the international meetings to arise from the Atoms for Peace program, the U.N. International Conference on Peaceful Uses of the Atom, first held in Geneva in August 1955. Its purpose was to provide scientists from both sides of the Cold

War the opportunity to meet for the first time since the onset of hostilities and to exchange ideas and information on the peaceful uses of the atom. The conference helped to reduce political tensions and public fears about nuclear war, and it helped to stimulate international cooperation through the IAEA, which is still active today in monitoring nuclear proliferation. But it also provided the Americans with new insights into Soviet science that, if anything, increased their fears about Soviet capabilities.

Far from the American stereotype of Soviet scientists as incompetent ideologues, Soviet physicists turned out to be highly capable researchers who delighted in the respect shown them by their Western colleagues. In addition to being impressed, their colleagues were shocked to learn of Soviet plans to challenge the American lead in one of the most prized symbols of scientific superiority, accelerator physics. The Soviets were planning the construction of a huge machine capable of producing particle energies reaching 10 billion electron volts, far in excess of the energies then attained by the biggest American machines.

Noting that "the Soviet Union has challenged our leadership," physicist Frederick Seitz leapt into action. He demanded that the Department of Defense, which provided 74 percent of federal research funds that year, establish a new high-energy physics funding program in addition to the program already funded by the AEC. It is "essential that the United States retain its leadership in all essential parts of the field," the solid-state physicist informed the military leaders.[23] Defense of America's lead in this field in particular was so important that, in his view, it was a potential matter of military concern.

MOBILIZING MALES

As federal funds flooded into domestic research after the war, American scientists became not only the best funded of the industrialized world, but also the most numerous—further enhancing their influence on the world stage. The demand for physicists in postwar industry and academe rose sharply, and with it the number of new PhDs. The Great Depression was over, demand for consumer goods was up, and the GI Bill was sending an army of war veterans to colleges and universities. The annual production of new physics PhDs jumped from 47 in 1945 to 500 in 1950, and it remained there approximately until the end of the 1950s.[24] According to data compiled by Rossiter (see Table 3 in the Appendix), by the mid-1950s the number of physicists and astronomers in the United States was 11,452. Of these it is estimated that one fourth were in academia and three fourths in industry. An analysis of the institutions producing physics PhDs during the 1950s reveals that the same top ten institutions that produced the most PhDs during the 1930s (accounting for 50 percent of all physics PhDs) were also among the top ten during the 1950s (now producing 40 percent of all PhDs). It was no accident that in both decades they were also among the largest recipients of federal and corporate funding. During the 1950s the top two PhD-producing institutions were UC Berkeley and MIT, and each produced nearly twice as many physics doctorates (339 and 303, respectively) as each of the other institutions.[25]

The demand for new doctorates and the flow of money into research also had a significant impact on the demographic makeup of the postwar physics profession. The war had brought large numbers of women, as substitutes for male workers, into broad areas of the nation's economy. During the early postwar years, many women, including those working in scientific fields, remained temporarily in place while the returning veterans, taking advantage of the GI Bill, obtained further education. The onset of the Cold War, beginning with the Berlin Blockade in 1948, followed by the outbreak of the Korean War in 1950, reinforced the government's policy of maintaining the nation's economy and its science on a permanent war footing. With the universal military draft of men still in place and a monthly quota of new inductees reaching 50,000 per month in 1950, a year later President Truman called for the establishment of a permanent standing army of 3.5 million men.

In view of the demands for both soldiers and scientists, the nation could not afford to ignore its underutilized human resources: women and minorities. The Office of Defense Mobilization (ODM), located in the President's executive office, called for mobilizing women and minorities for careers in science and engineering (though not for military service) and for employers to employ them. The NSF and ODM undertook a number of statistical studies of the nation's scientific manpower, of great value to us today, and a host of articles in science publications called for increased training of female scientists and engineers. More access to science for women and minorities remained the government's official policy for nearly a decade. An ODM report in 1952 recommended that employers "re-examine their personnel policies and effect any changes necessary to assure full utilization of women and members of minority groups having scientific and engineering training."[26]

Nevertheless, Margaret Rossiter has found that the new policy remained little more than empty rhetoric. There were no federal incentives or enforcement to back up the federal recommendations. By 1954–1955 women were being encouraged to enter science teaching rather than science research. By 1960 opposition to the recommendations had already mounted.[27]

Rossiter has compiled a wealth of information and statistics from the NSF and other "manpower" studies that portray once again the continuing underrepresentation of women in physics and other sciences during the decades following World War II, despite the obvious need for their participation in the nation's science and engineering. (Data on minorities are not currently available.) Table 3 in the Appendix displays the steady growth in the number of scientists and engineers and in the number of physicists and astronomers in the United States from 1955 to 1970 (data for these disciplines were combined). Much of this occurred, of course, in reaction to the launch of the Soviet *Sputnik* satellite in 1957. The number of scientists/engineers and physicists/astronomers roughly tripled during that period, as did the number of men in each category. But even though the number of women in both categories roughly quadrupled, the increase in their percentage representation was far more modest. The representation of women physicists and astronomers during the massive build up of science and engineering after the war, then again after *Sputnik*, still remained at about the same percentage of the profession in 1970 as it had been in 1938 and even in 1921! (Compare Table 1 in the Appendix.) With the prevalent stifling stereotypes of family and gender during the 1950s, women were still discouraged from entering science and were apparently still considered unsuited for scientific research.

Rossiter showed further that, despite the overall burst in the number of science doctorates during the period from 1948 to 1961, female scientists received only about 8 percent of the doctorates awarded in science. In the Purdue University chemistry department, for instance, the number of women students and their choice of specialties were severely limited because so few professors were willing to take on a female student. Only 115 of the over 6,000 physics doctorates in that period went to women, a mere 1.87 percent. The most popular doctoral fields for women were psychology, biosciences, and even chemistry, but women were still vastly underrepresented in those fields as well. While, as noted earlier, the University of California at Berkeley and MIT awarded by far the most PhDs in physics, only Berkeley (appearing together with UCLA) is in the list compiled by Rossiter of the top twenty-five institutions awarding science doctorates to women, including doctorates in "physics and meteorology."[28]

The employment picture reflected a parallel underrepresentation of women physicists in both industry and academe. Even as the numbers of male and female physicists in industry increased during the years 1958 to 1968, female representation still dropped over that decade from 1.51 percent to just 1.41 percent. By 1970 about 38 percent of women scientists employed in industry or academe were engaged in teaching, about a third were employed in research, and only about 9 percent worked in management.[29]

The early decades of the Cold War and the efforts to mobilize science and scientists in defense of the nation and its culture clearly did not extend to women, and least of all to African Americans and other minorities. In 1973 Shirley Ann Jackson graduated from MIT, the first African-American female doctorate in physics. Herman Russell Branson was one of the first male African-American doctorates in physics when he graduated from the University of Cincinnati in 1939. Despite the nation's needs and the government's recommendations, women scientists achieved only minimal gains in the still largely white-male dominated disciplines of physics and of science in general.

6
The New Physics

World War II and the postwar aftermath brought striking changes to the structure of the physics discipline and to the nature of its work. Not only did the high demand for physicists during the war continue after the war, but also, with money flowing, big projects, big teams, and big budgets became common—a reflection in many ways of the mass production characteristic of the postwar consumer society. As funding increased the number of researchers and the outpouring of their research, the lone researcher tinkering in a laboratory or sequestered in an office with a pad of paper and a pencil had nearly become a thing of the past. Nevertheless, a number of individual researchers and small groups of researchers did manage to make important breakthroughs in this period. Owing to the practical needs of the military and industrial funders of research, many of these breakthroughs occurred on the border of science and technology, as Paul Forman has argued. Yet, even as individuals, their discoveries would not have been possible without the large-scale wartime and postwar research that preceded their work, or the military resources that funded theirs and prior work, or the huge institutional organizations that supported them. As Spencer Weart writes, "American physics was no longer like a small town where everyone knew each other."[1] Now in a big city, some physicists tended to identify more with their local neighborhoods: their individual subdisciplines and local research teams.

Through its wartime contributions, theoretical physics had gained by now a status comparable to experimental research. Nevertheless, most of the dividends of postwar research accrued instead to experimental physics as the result of the wartime successes.[2] Those dividends and the impetus of the MIT Radiation Laboratory also enabled the emergence of a new physics discipline that drew adherents from a number of related disciplines. As early as 1944 the American Physical Society established the Division for Solid-State Physics, a division that drew together academic and industrial physicists in the study of the properties of solids, their quantum origins, and their practical uses. (Today, with the inclusion of liquids, the field is known as condensed-matter physics.) Spencer Weart has described how this field emerged and established itself by 1960—with the direct help of the Office of Naval Research (ONR) and the Department of Defense—as an independent discipline with its own journals, textbooks, meetings, prizes, and buildings, first in the United States and then in Europe and abroad.[3]

"THE CRYSTAL MAZE"

Following the development of microwave radar during the war, a number of university and industrial laboratories began investigating semiconductor crystals after the war as the basis for developing smaller and more efficient electronic components to replace vacuum tubes. They already had experience during the war with one solid-state component, a crystal diode rectifier able to withstand the high "back voltages" required for the detection of microwave radar beams (as the standard radio-wave vacuum tubes of the day became unstable at these voltages). In 1942 the MIT Radiation Laboratory had awarded an Office of Scientific Research and Development subcontract to the Purdue University physics department for supplemental work on the development of a suitable diode rectifier. Austrian born and educated department head Karl Lark-Horovitz had gained experience with crystal-diode radio during World War I, but vacuum tubes had long since replaced diodes in radio technology. While the MIT Rad Lab investigated silicon, the Purdue team under Lark-Horovitz worked in a different direction—toward the investigation of the still little-researched element germanium as the basis for the needed rectifier. Germanium and silicon are normally insulators, but researchers at the Sperry Gyroscope Company had gained the first evidence that by "doping" germanium and silicon with certain impurities they could create so-called n-type and p-type "semiconductors" that can transmit currents of either negative or positive charges, respectively. Because of this, the electrical properties could be easily controlled and used for a variety of purposes. Purdue physicists undertook careful studies of this effect in germanium. By spring 1943 they had produced the first high back-voltage germanium diode. The MIT Rad Lab assigned mass production of this early device to Bell Telephone Laboratories.[4]

The Purdue team continued to perfect high back-voltage germanium rectifiers, contributing to the development of the more commonly utilized version, and they investigated theoretically and experimentally the electronic properties of germanium in general. By the end of the war, the department was even producing its own stock of high-purity germanium-crystal ingots for further research. Theorist Vivian A. Johnson, who played a key role in the Purdue germanium research, reported that at the end of the war the group decided "to abandon development of detectors and the practical applications and to concentrate on the basic investigation of germanium semiconductors." Following the declassification of their work, the Purdue physics department found itself on the forefront of solid-state research. As other universities and laboratories joined the work on germanium and other semiconductors, in January 1946 Lark-Horovitz reported for the first time on the physics of germanium to an overflow audience at the annual meeting of the American Physical Society. The Purdue work played a vital role not only in advancing solid-state research but also in a discovery soon to emerge from Bell Telephone Laboratories.

In 1945 Bell Laboratories became the first and for many years the only institution to establish an entire department dedicated solely to the study of solid-state physics. Mervin Kelly, the director of research, stated that with the utilization of quantum mechanics, "a unified approach to all of our solid state problems offers great promise."[5] One of the solid-state problems for the new department was to move beyond crystal diodes to the development of a solid-state triode amplifier to replace the commonly used vacuum-tube amplifiers still used in long-distance telephone service. Within the new department, William Shockley, a physics graduate of Caltech and MIT, headed the semiconductor division, which included John Bardeen and Walter Brattain. Shockley and Brattain, adept in experimental solid-state physics, had worked during the war at Columbia University on submarine detection under contract to the Office of Scientific Research and Development. Bardeen, a theoretical physicist from Princeton, had worked at the Naval Ordnance Laboratory during the war. Because of Purdue's pathbreaking work on germanium, they focused on germanium rather than silicon as the basis for their new device. Utilizing quantum theory and doped germanium crystals supplied by Purdue, their research led Bardeen and Brattain in 1947 to the discovery, briefly put, that a small electric current emitted onto the surface of a germanium crystal diode is amplified after passing through the semiconductor. The device was called the point-contact "transistor," named because it transmitted or resisted current depending on the voltage at the base of the crystal. The new device and its later improvements enabled the replacement of vacuum tubes by much more reliable, smaller, more energy-efficient solid-state electronics. With Shockley's participation, in 1951 the team joined together three doped germanium crystals into what became today's triode "junction transistor." It proved more mechanically stable and much easier to manufacture than the point-contact device. Shockley, Bardeen, and Brattain received the 1956 Nobel Prize for physics "for their researches on semiconductors and their discovery of the transistor effect."[6]

The transistor could be used not only as an electronic amplifier but also, through its off and on states controlled by the voltage on its base, as a switching device of use in telephone connections as well as logic circuits. The off and on states provided an electronic representation of the digital binary numbers 0 and 1.

In 1959 Jack Kilby, an engineer at Texas Instruments, constructed the first transistor-based circuit on a single wafer of germanium crystal. A year later, Robert Noyce, a physicist at Fairchild Semiconductor, invented the photolithography technique for producing electronic microcircuits on a semiconductor crystal base. This process became the standard technique for manufacturing "integrated circuits," circuits that combine huge numbers of microscopic transistors and other electronic components on a single crystal wafer. Integrated circuits quickly became the backbone of modern electronics, contributing to the invention of ever smaller, rugged, and more energy efficient telecommunications, entertainment, and sensing devices, and helping to unleash the computer and digital revolutions at the foundation of today's "information age."[7]

Kilby received the 2000 Nobel Prize in physics for his contribution to the invention of integrated circuits, despite its applied nature. Noyce and Gordon Moore cofounded the Intel Corporation in 1968. Without the breakthroughs in transistor technology and integrated circuits arising from prior large-scale fundamental physics research, the computer and digital revolutions, and the many digital devices they spawned, would not have been possible.

FROM MASERS TO MAGNETIC RESONANCE

Physicist Charles H. Townes was another example of an individual researcher whose work and breakthroughs were possible because of, rather than in spite of, large-scale, largely military funded team research. Townes had received his doctorate at Caltech in 1939 with a dissertation on isotope separation, but he spent the war years working at Bell Labs designing radar-based bombing systems for the army. In 1948 he joined the faculty of Columbia University, where he applied the microwaves of radar beams to the study of the structure of molecules. Some molecules absorb and emit microwaves in distinctive spectroscopic patterns that reveal their chemical structure. Funded by the Office of Naval Research (ONR), in 1951 Townes conceived of utilizing a quantum process, published by Einstein in 1917 before the advent of quantum mechanics, in an effort to generate more powerful beams of microwaves for use in improved radar systems. It was an idea also pursued by Soviet scientists at that time. Utilizing quantum theory, Einstein had shown that photons passing through a gas will stimulate the atoms of the gas to emit additional photons of precisely the same frequency. Using microwaves passing through a device filled with ammonia gas, Townes and his collaborators achieved the first stimulated emission of electromagnetic radiation in 1954. He called his device the "maser," for "*m*icrowave *a*mplification through *s*timulated *e*mission of *r*adiation."

If microwaves could be amplified into a radar beam, then perhaps even shorter wavelength waves such as infrared or even visible light might also be amplified though stimulated emission. Townes and his students, along with physicists in other laboratories, began working toward building such a device. When in 1958 Townes and his brother-in-law Arthur Schawlow published the theoretical design of a visible-light maser using mirrors at each end of the cavity, the race was on to actually build one. In 1960 Theodore Maiman at Hughes Aircraft Laboratories in Culver City, California, crossed the finish line first when he sent pulses of red light through a crystalline pink ruby and attained the first stimulated emission of visible light. It was the world's first "laser," where *l* is for light.

What made the laser so significant was that the amplified light was totally coherent, that is, all of the waves are identical and oscillate precisely in phase with each other, allowing the beam, unlike normal light, to be focused on an extremely narrow point and with little dissipation of the beam over large distances. The laser has proved to be indispensable in the laboratory for many types of experiments. Like the transistor, the laser has also enabled a wide range of commercial applications that have transformed our lives, from audio and video entertainment (CDs, DVDs, Blu-ray discs) to data handling, manufacturing processes, and medical applications that would not have possible without the breakthroughs in fundamental physics behind them.[8]

In 1945 a simultaneous discovery—the first observation of nuclear magnetic resonance in bulk matter—exerted an equally profound impact on postwar research in a wide variety of fields and in medical diagnosis. Most atomic nuclei are magnetic, that is, they behave as tiny magnets. Their magnetic strength is represented by the so-called magnetic moment of the nucleus. If the nuclei are subjected to a strong, constant outside magnetic field, they will orient themselves in the direction of the field, just as does a compass needle in Earth's magnetic field. However, because of quantum mechanics, they will not orient their axes exactly along the field but at an angle, while the axis of each nucleus rotates around the field direction at a frequency determined by quantum theory. As noted earlier,

Isidor I. Rabi had predicted and confirmed in 1938 that he could cause the nuclei in a magnetic field to undergo what is called resonance. He passed a beam of identical atoms through a strong magnetic field. He then applied a smaller oscillating magnetic field at right angles to the first field and adjusted the oscillation of this field until it reached the rotation frequency of the nuclei in the initial field. At that instant, resonance occurred, as evidenced by the absorption of energy from the oscillating field and the reduction in its intensity. From this "nuclear magnetic resonance," or NMR, Rabi and coworkers were then able to determine the magnetic moments of the nuclei.

Edward Purcell had been working during the war at the MIT Rad Lab where he had observed the absorption of radio frequency energy by matter at certain resonance frequencies. One year after Rabi received the Nobel Prize for this work in 1944, Purcell, Henry Torrey, and Robert Pound, working at the MIT Rad Lab, investigated the radio frequency magnetic resonance of nuclei in the molecules of solid paraffin wax, a material that is rich in carbon and hydrogen. At the same time, on the West Coast at Stanford University, Felix Bloch. W. W. Hansen, and M. Packard independently undertook a very similar study. Their purpose was to observe NMR in a solid, as Rabi had done for free atoms in a beam, and to identify the magnetic quantum energy levels of the nuclei by detecting the resonance transition between levels. Both groups succeeded and submitted papers to *Physical Review* within a month of each other. Purcell and Bloch received Nobel Prizes for the work in 1952.

Nuclear magnetic resonance was extremely useful for measuring the magnetic moments of the nuclei of all elements and in solid, liquid, and gaseous states. But soon it proved even more useful in identifying unknown nuclei and elements and their arrangement in a piece of matter. This effort was soon aided in 1953 when Albert Overhauser, later a Purdue professor, proposed the idea that a small alteration in the alignment, or polarization, of the electrons surrounding the nuclei in the atoms and molecules of the material will greatly enhance the polarization of the nuclei in an external magnetic field by 1,000 times or more, thereby making the resonance even more obvious.

Confirmed shortly thereafter by T. R. Carver and C. P. Slichter, the Overhauser Effect became a crucial tool in the NMR determination of complex molecular structure in chemistry, and it has aided in the application of NMR to medical imaging in the form of magnetic resonance imaging (MRI). Nuclear magnetic resonance has found wide applications in chemistry, nuclear physics, geophysics, and even quantum computing. With the advent of radio astronomy after the war, Purcell himself brought NMR spectroscopy to astrophysics, using resonances detected through radio astronomy to investigate the interaction of interstellar dust with light propagating through the galaxy.[9]

As with transistors, lasers, and masers, NMR arose not only from its favorable financial and institutional environment but also from the discovery of an effect based upon fundamental quantum principles that then led to a wide range of practical applications and to many research applications linking together a wide variety of disciplines.

RENORMALIZING

In addition to its role in the development of transistors, lasers, and NMR spectroscopy, the radar work of the MIT Rad Lab during the war years also played an important role in the drama of postwar theoretical physics. Before the war, two popular areas of theoretical research, quantum electrodynamics and theoretical high-energy physics, had stalled upon encountering seemingly insurmountable obstacles. The obstacles were finally surmounted during a series of small, federally funded conferences held soon after the war, igniting a burst of new research into high-energy physics and quantum field theories that helped maintain American physics at the forefront in these areas.

For three years starting in 1947, the National Academy of Sciences sponsored a series of annual conferences in isolated vacation resorts in New York State on the theoretical physics of quanta and fields. These conferences, called the Shelter Island conferences after the location of the first meeting, played a crucial role in stimulating postwar high-energy physics. They led to the solution of the lingering problems that had been hindering progress, and they elevated some of the brightest young physicists to come out of war research into the highest echelons of this prestigious specialty. Among the attendees at the initial conferences, chaired by the ever present J. Robert Oppenheimer, were Richard P. Feynman, Julian

Schwinger, Freeman Dyson, Abraham Pais, and Robert Marshak. More-over, these conferences, writes Schweber, "reasserted the values of pure research and helped to purify and revitalize the theoretical physics community" after its wartime contributions to the atomic bomb.[10]

As noted in Chapter 3, one of the biggest problems facing relativistic quantum electrodynamics was the appearance of infinite results in the calculation of quantum energies. These included even the so-called self-energy of electrons, such as those orbiting the nucleus in an atom. The second puzzle concerned the penetration of cosmic ray particles, mesons, through matter. Mesons were predicted by Yukawa's quantum theory of the strong nuclear force. They are created naturally when high-speed cosmic rays from outer space collide with atoms in the upper atmosphere at energies well above 100 million electron volts (MeV). Once they reach the ground, the high-energy mesons should not penetrate very far into a lead plate in a laboratory experiment, but in fact they do, in stark contrast to theoretical predictions. Did the theory simply break down at high energies?

During the first of the National Academy conferences, held on Shelter Island on the eastern end of Long Island, Willis Lamb, a former Oppen-heimer student, opened the proceedings with puzzling data on the spectroscopic lines emitted by hydrogen. Utilizing microwave radar technology developed during the war, Lamb and his student Robert Retherford at Columbia University had found that one of the quantum energy levels in the hydrogen atom lay slightly higher in energy than a nearby energy level. This was the so-called Lamb shift. It conflicted, however, with the prediction of quantum electrodynamics that the two levels should have the exact same energy. In addition, Bruno Rossi, who had arrived in Chicago from Rome with Enrico Fermi during the late 1930s, reported on new cosmic ray experiments confirming that high-energy mesons do indeed penetrate through matter much farther than quantum theories allow.

With input from Japanese physicist Sin-itiro Tomonaga via letter, a solution to the first puzzle gradually emerged during the next conferences. Schwinger and Feynman demonstrated that quantum electrodynamics could account for the Lamb shift if the infinities were removed from the mathematics through a process called "renormalization." In this process, certain fundamental parameters, such as the charge of the electron, are simply redefined so as to cancel out the infinity from the calculation, leaving behind a finite result for comparison with the data. Aside from questions about the acceptability of this procedure, it still had to be rendered fully compatible with the demands of relativity theory. Schwinger managed to achieve the needed compatibility through a highly mathematical analysis, while Feynman offered an equivalent alternative approach involving a new and powerful visualization of interactions of particles and fields now known as the famous "Feynman diagrams." They have provided ever since an elegant and efficient means of calculating observable quantities such as the Lamb shift. Freeman Dyson soon proved that the renormalization of electrodynamics could be extended to the relativistic quantum mechanics of other fields, such as Yukawa's meson field or Fermi's weak nuclear force. After decades of encountering insurmountable infinities in their calculations, theorists were now able to compute experimentally verifiable atomic and quantum electromagnetic phenomena involving point-sized electrons and the very highest energies.[11]

The second leftover puzzle—meson absorption—also began to succumb to the conference investigators. Marshak, later publishing with Hans Bethe, offered the brilliant suggestion that perhaps there are actually two different mesons arising from cosmic rays impinging on the atmosphere. One of these, Yukawa's meson (now called the pi-meson, or pion), is produced high in the upper atmosphere; the other (now called the mu-meson, or muon) was envisioned as a product of the decay of the Yukawa meson. This second meson would interact much less readily with matter, leading to the observed penetration of lead blocks at sea level.

Subsequent experimental investigations on cosmic rays strikingly confirmed Marshak's hypothesis. At the same time, the Groves-funded synchrotron at Berkeley was already surpassing 200 MeV. By 1948 it was producing both types of Marshak's mesons. High-energy mesons and their interactions could now be artificiality produced and studied in the laboratory, rather than studied solely in difficult cosmic ray experiments. Physicists now had even stronger reasons to induce the Atomic Energy Commission (AEC) to build ever more and bigger accelerators and the laboratories to house them.

With renormalized theories working hand-in-hand with improved experimental techniques, theoretical research on quantum field theories received new impetus, while at the same time high-energy physics, employing large teams of researchers working on expensive machines, emerged at last from nuclear physics as an independent discipline.[12]

PUTTING PHYSICS IN ITS PLACE

With the establishment of the ONR, the AEC, and the National Science Foundation, American physicists had achieved, in the words of Daniel Kevles, what they had sought since the days of Henry Rowland: "a system of federal support for basic research and training insulated from political control and focused on the advancement of the best possible physics."[13] Although the insulation from military control was far less assured, this system made possible the big advances outlined above that the profession had achieved during the postwar period. But it also rendered even more acute the fundamental question about pure science that had arisen at various times in the past. In the words of policymaker Don K. Price in his influential book of 1965, *The Scientific Estate*, "How is science, with all of its new power, to be related to our political purposes and values, and to our economic and constitutional system? By ignoring this question, we have been trying to escape to science as an endless frontier, and to turn our backs on the more difficult problems that it has produced."[14] While that question did not escape attention beginning in the 1960s, the related question of how much political, ethical, and even moral responsibility to allow physicists to exercise regarding the use of their work had already arisen with Bush's handling of wartime policy. It now required a more immediate answer as Cold War tensions increased during the 1950s. Should scientists be granted a responsible role in policy decisions, or should they be limited to providing only technical expertise? Or, as it was put more pithily, should scientists be "on top or on tap"?[15]

The earlier reactions of many physicists to the May-Johnson Bill for the AEC already hinted at *their* early views on these matters. While most federal officials looked upon physics as primarily an instrument of national security, many of the physicists who built and helped use the atomic bombs began to express quite different concerns regarding nuclear weapons after the war ended. After all, as the informed builders of the bombs that had been dropped on Japan, they felt they should have a voice in the direction of future development and use of such a weapon. In a remarkable outpouring of sentiment, 515 physicists meeting in Cambridge, Massachusetts, less than three months after the atomic bombing of Japan signed a petition calling for the end of all war and for the international control and eventual elimination of all nuclear weapons.[16] It went nowhere. At the same time, Oppenheimer, then still head of the Manhattan Project and still chair of the Interim Committee's Scientific Advisory Panel to the War Department, presented a letter to the War Department (soon renamed the Defense Department) urging restraint and the search for the international control of nuclear weapons. However, as he reported to Ernest Lawrence, the attitude at the War Department was "full steam ahead."[17]

The developing international situation did not help matters. The failure of the Baruch Plan, a U.S. proposal submitted to the U.N. in 1946 for the international control of nuclear weapons, marked an early impetus to the nuclear arms race. As the race gained momentum, tensions mounted during the late 1940s when the Iron Curtain fell across Soviet-dominated Eastern Europe and the Russians set in place the Berlin Blockade. The world sank into a Cold War between the United States and the Soviet Union and their respective allies. While fear and hysteria gripped segments of American society and military funds flowed freely into physical science, physicists were torn between an obligation to defend their country and their desire to avoid building ever more and bigger bombs, all under the constraints of military secrecy and control.

As he did at Los Alamos, Oppenheimer came to represent to the general public and to physicists alike the new status of the physics profession in the upper echelons of the nation's centers of power and policymaking. Although the battles over the dropping of the atomic bomb and the founding of the AEC had shaken physicists' trust in his role as one of their main representatives in government, Oppenheimer and his closest colleagues were still held in high regard.

That assessment was put to the test beginning on September 23, 1949. On that date, President Truman announced to a stunned nation that long-range sensors had detected evidence that the Soviet Union had detonated its first atomic bomb.[18] The bomb had been tested on August 29, 1949— far earlier than anyone had expected, owing in part, it was learned, to the work of Soviet spies on the Manhattan Project.

The news sent the nation's leaders into a panic, increasing the likeli-hood for them of what they feared would be "another Pearl Harbor." Within days of the news, Lawrence and the MIT Rad Lab's vice director, Luis Alvarez, descended upon Congress and the Pentagon in a full-scale lobbying effort for a crash program to build the next generation of nu-clear weapons, the "super" or hydrogen bomb. During his youth, Edward Teller had experienced a Communist regime that had gained temporary power in his native Hungary.[19] With first-hand experience of a Commu-nist dictatorship and worried about the consequences of a Soviet victory in a nuclear war, on October 13 Teller submitted a memo to the Los Ala-mos leadership arguing that peace with the Russians would be possible only if the United States possessed "overwhelming superiority." Now that the Russians had an A-bomb, it was assumed they would surely proceed to the next level of nuclear weaponry, the H-bomb—even if no one yet knew how to build such a device. In his only television appearance, Einstein called Teller's argument that the building of ever bigger nuclear weapons would maintain peace a "mechanistic, technical-military, psychological attitude."[20]

The principle behind the hydrogen, or fusion, bomb had been well known to researchers even before the war. While the "atomic" bomb de-rived its power from the splitting, or fission, of the heavy nuclei of ura-nium and plutonium atoms, the hydrogen bomb derived its energy from the opposite process, the fusing together of two light nuclei, such as hy-drogen, to form a heavier nucleus through the action of the short-ranged strong, or nuclear, force. The predicted energy released in this process would be at least 1,000 times greater than that released in a comparable fission bomb. Its ultimate power was limited only by the amount of fusion material available.

Fortunately, the fusion process is extremely difficult to initiate, because only temperatures in excess of 1 million degrees can bring the electrically repelling nuclei close enough together to trigger a fusion reaction, and then only under special conditions. These enormous temperatures occur within the sun and other stars, which are powered by fusion reactions. But on Earth they were, at that time, attained in only one way—at the center of an atomic bomb blast. Perhaps a way could be found to use a fission bomb as the trigger for a fusion bomb. Because of the high temperatures involved, this type of bomb (if it was small enough to be dropped as a bomb from a plane) was also called a "thermonuclear device."

During and after the war, Edward Teller had pressed for larger scale work on the H-bomb, but Oppenheimer and his committees had thwarted those plans on every occasion. The veterans of Los Alamos were still shaken by the devastating effects of the fission bomb, and the likelihood of achieving a fusion bomb any time soon seemed very remote. But now, in the wake of the Soviet atomic bomb, on October 11, 1949, David Lilienthal, director of the AEC, asked the AEC's General Advisory Committee to advise on the feasibility and necessity of a crash program to build the H-bomb and on broad policy regarding fusion weapons. After intensive deliberation under Oppenheimer's chairmanship, the eight members of the civilian committee (Seaborg was absent) were unanimous in opposing a crash program to build the H-bomb.

Their decision rested on narrow technical concerns as well as on broad considerations of military policy (as requested) and even on moral considerations. They were convinced that the Russians would not soon have an H-bomb, that all current plans to build such a device would not work, and that American conventional forces and the current arsenal of fission bombs were sufficient to defend the nation against any attack. In addition, they were uncomfortable about recommending another Manhattan Project. "We already built one Frankenstein," one said.[21] But even more discomforting, the H-bomb represented to them a quantum leap in nuclear destruction. Its sheer, potentially unlimited power meant that its primary use would not be for tactical battlefield combat but for the destruction of whole metropolitan areas. "Its use therefore carries much further than the atomic bomb itself the policy of exterminating civilian populations," the committee declared.[22]

This conclusion induced the scientists to go beyond technical and military policy considerations and to raise the moral issue involved in the use of a weapon of such enormous destructive power. Six of the scientists warned in an appendix to the General Advisory Committee report stating that "a super bomb might become a weapon of genocide." Fermi and Rabi declared in a separate appendix, "Its use would put the United States in a bad moral position relative to the peoples of the world . . . It is necessarily an evil thing considered in any light."[23] But Teller responded by reasserting the wartime view that physicists should stick to technical details and leave the larger issues to their superiors: "It is not the scientist's job to determine whether a hydrogen bomb should be constructed, whether it should be used, or how it should be used. This responsibility rests with the American people and their chosen representatives."[24]

On January 31, 1950, the people's chosen representative, President Harry S. Truman, ordered continued work on the hydrogen bomb over the objections of most physicists. Five days after Truman's secret order, twelve prominent physicists, among them Hans Bethe, George Pegram, and Victor Weisskopf, urged the president on moral grounds to issue a second order declaring that the United States would never make first use of such a weapon. "This bomb is no longer a weapon of war but a means of extermination of whole populations. Its use would be a betrayal of all standards of morality and of Christian civilization itself," they wrote.[25] For tactical reasons, Truman refused to renounce the first use of any nuclear weapons, as have all presidents ever since.

Two months after Truman's order, the revelation of a Soviet spy at Los Alamos, Klaus Fuchs, who was privy to H-bomb work, induced Truman to order not just continued work, but a crash program to build the new bomb. Hans Bethe and others now reluctantly agreed to join work on the bomb in order, he later explained, to maintain "the balance of terror."[26] One side effect of this work was to stimulate research in the field of plasma physics, including the physics of stellar evolution. This work led to a return to the problem of the gravitational collapse of a star into a singularity, what was later called a black hole. This was a phenomenon first discovered by Oppenheimer and his colleagues in theoretical terms during the late 1930s but set aside at the time, apparently because of the difficulty of dealing with mathematical infinities when other phenomena were of equal or more interest at the time.[27] (Black holes are further discussed in Chapter 8.)

The design of the H-bomb required the heat of a fission bomb to trigger the hydrogen fusion reaction. The hydrogen isotopes deuterium and tritium were used as fuel sources. But a design by Teller in which a fission device was placed next to a container of liquid deuterium failed to fuse because the liquid did not compress. In 1951 Stanislaw Ulam, a Polish-born mathematician, devised a process called staging, whereby the fission explosion was designed to compress the liquid in addition to heating it.

Teller added the innovation that the electromagnetic radiation emitted by the fission bomb, not the shockwave, could be used for this purpose. In addition, a "spark plug" of plutonium was embedded inside the deuterium container so that it would implode along with the compressing deuterium, enhancing the likelihood of achieving the much more powerful fusion explosion.

However much he detested the device, physicist Oppenheimer called the design "technically sweet."[28] Using the Ulam-Teller design, on November 1, 1952, the United States detonated the world's first thermonuclear device on the Eniwetok Atoll in the South Pacific. The device (too big to drop as a bomb) released an unimaginable amount of energy equivalent to 10 million tons of TNT, obliterating the atoll. In comparison, the fission bombs dropped on Hiroshima and Nagasaki were equivalent to about 12,000 to 20,000 tons of TNT.[29]

Although Oppenheimer, DuBridge, Conant, and other opponents of the H-bomb continued to advise the military, Teller and those who supported the new bomb now gained greater influence in policy matters. At the same time, the Cold War grew ever colder and ever more fearful. The Korean War nearly led to nuclear war. As Senator Joseph McCarthy and his cohorts fueled the "red scare," the Soviet test of a prototype H-bomb struck additional fear across the political landscape. When Oppenheimer again advised against the use of the H-bomb for strategic offense against the Soviet Union, General Curtis LeMay, commander of Air Force's Strategic Air Command, which was preparing to deliver the bombs over Russia, flew into a rage. It was time to make an example of Dr. J. Robert Oppenheimer.

An anonymous article in *Fortune* magazine accused Oppenheimer of having purposely delayed the hydrogen bomb.[30] It was long known to intelligence agents that Oppenheimer had associated with Communists in Berkeley during the 1930s, but there was no clear evidence of disloyalty, nor had any arisen since. The new director of the AEC, Lewis Strauss, was personally opposed to Oppenheimer, and, with President Eisenhower unlikely to defend Oppenheimer, the AEC's general manager, Major K. D. Nichols, presented Oppenheimer with a bill of charges in late 1953. He accused Oppenheimer of delaying the hydrogen bomb, of persuading others to do likewise on nonscientific moral grounds, and of having doubtful explanations about his prior associations with known Communists. All of these claims, wrote Nichols, "raise questions about your veracity, conduct and even your loyalty."[31]

The AEC suspended Oppenheimer's security clearance. Without it he could not provide any further advice to the government on classified matters. Strauss gave him the choice of quietly resigning from the AEC or of contesting the accusations in a full-scale hearing. Oppenheimer decided to contest the charges, setting in motion a lengthy hearing that painfully delved into nearly every aspect of his life and work.

Numerous books, articles, and even a popular play have been written about the Oppenheimer case from many different perspectives.[32] Oppenheimer had made serious missteps regarding his earlier Communist associations, in addition to antagonizing powerful people, but the proceedings were not exclusively about solving a security risk. Oppenheimer was not a consultant at that time, and his security clearance was set to expire anyway, one day after the hearings ended. As Rabi noted, "if you don't want to consult the guy, don't consult him, period."[33] Rather, in the depths of the Cold War, which threatened to become a nuclear hot war at any moment, the national-security state decided to make an example of one of its most prominent leaders. The message was that, as in the wartime Manhattan Project, physicists should be on tap, not on top. They were to stick to technical details and avoid expressing any troubling concerns or contributing any broader policy statements regarding their work and its uses unless their views supported those of their superiors.

At the end of May 1954, the AEC's three-man Personnel Security Board found, by a vote of two to one, that Oppenheimer "is a loyal citizen" but nevertheless "a security risk." The lone dissenter was the only scientist on the board, a chemist. The main reasons given by the majority were his "attitudes": first, his lack of "enthusiastic support for the [H-bomb] program"; then, his audacity in expressing convictions as a "scientific advisor" that "were not necessarily a reflection of technical judgment, and also not necessarily related to the protection of the strongest offensive military interests of the country." It was no matter that Oppenheimer was one of only eight members of a committee that was unanimous in its convictions. The judgment concluded: "Any man, whether specialist or layman, of course, must have the right to express deep moral convictions . . . [But] emotional involvement in the current crisis, like all other things, must yield to the security of the nation."[34]

Oppenheimer, the hero of the Manhattan Project who had continued to serve his country for the past nine years, was removed as a government advisor and sent back to Princeton in disgrace.

The Oppenheimer case made clear to all the political and social realities of physicists as partners in government-sponsored research. The grand visions of scientists as participants in military policymaking or responsible moral authorities faded against the fears for national security in the midst of the deepening Cold War. Even the public's heroic image of scientists as the winners of the world war had faded after nearly a decade of nuclear Cold War. "So you think a scientist is a 'sacred cow'!" one irate woman wrote to Caltech president Lee DuBridge who had publicly defended Oppenheimer. "Well, many Americans do *not*." Another woman wrote, "The American people are getting fed-up on the over-emphasis of 'science.' The most recent and most spectacular contribution to science is a creation that so far has brought the world nothing but fear and destruction."[35]

Numerous physicists at national laboratories and universities signed petitions to Strauss and Eisenhower warning of the potential damage to AEC programs. Troubled by the rising threat of resignations, Strauss toured the weapons laboratories that summer to calm the simmering revolt. Although the scientists were angry, sad, and anxious about their future, Strauss was relieved to find no mass exodus from the laboratories. Nevertheless, the AEC had lost the complete confidence of the nation's scientists.[36]

With money still flowing freely in support of expensive projects, most physicists found ways to accommodate themselves to the new situation. At Berkeley, the big-machine particle physics of the Rad Lab continued to churn out new discoveries. As the top producer of physics PhDs, the campus, like others, was flooded with so many bright young physicists that most felt that losing Oppenheimer was no great loss. "As the Oppenheimer case showed so clearly," writes historian Paul Forman, "scientific talent was available in sufficient depth at all levels that an ungenial individual was a dispensable individual."[37] A decade after the war, physicists continued to enjoy lavish government funding motivated by the potential benefits of their work, but, for the time being, they were no longer treated as fully independent nor were they regarded as responsible equal partners with their military benefactors.

7
Sputnik: Action and Reaction

The man who presided over a nation confident of its century-defining stature in military, economic, and cultural matters could hardly believe what he heard on the evening of October 4, 1957. The president of the United States learned that evening that the Soviet Union had used the occasion of the International Geophysical Year, a period of international geophysical research of planet Earth, to launch into orbit the world's first artificial satellite. It was a sphere weighing about 184 pounds (84 kg), encircling Earth every 96 minutes at a speed of nearly 16,000 miles per hour while emitting a radio frequency beeping sound to announce its arrival. It was called *Sputnik*, Russian for satellite. Nearly a month later, the Russians outdid themselves with the launch of *Sputnik 2*. Weighing 1,100 pounds, it carried a live dog into orbit together with the equipment needed to sustain the dog and transmit its condition to ground controllers.

The unfortunate dog lived only a few hours, but the impact on the United States lasted over a decade. If the Russians could put satellites into orbit, they could also orbit nuclear weapons over the United States or use their powerful rockets to rain nuclear destruction on any point on the planet with less than an hour's notice. If a dog could be launched into orbit, so too could a human. Americans worried that soon the moon itself would become a Soviet colony. A humiliating American attempt on December 6 to launch a mere 4.4-pound (2 kg) satellite into orbit blew up on the launching pad. The United States finally succeeded with the launch into orbit of *Explorer 1* on February 1, 1958, but by then the lesson had already been learned. The nation's main competitor as a superpower had suddenly sprinted ahead in the prestigious, and threatening, fields of missile and space technology.

Before becoming president in 1952, Eisenhower, the supreme commander of the successful D-day invasion of France, had served after the war as the commander of the North Atlantic Treaty Organization (NATO) forces in Europe. This position had brought him into direct contact with those leading scientists who had opposed the crash program to build the hydrogen bomb in 1949. Eisenhower had agreed with the opposition physicists that conventional forces along with smaller tactical fission nuclear weapons were more suitable for repelling a Soviet ground invasion of Western Europe. But now, as the beeping *Sputniks* circled overhead, many of those who had opposed the H-bomb were no longer in government, and those who remained were preparing to disband their little-used advisory committee for lack of business. Eisenhower suddenly summoned its members to the Oval Office on October 15, 1957.

Chaired by physicist Isidor I. Rabi, the president's Science Advisory Committee (SAC) included Cornell physicist Hans Bethe; Detlev W. Bronk, the president of the National Academy of Sciences; James R. Killian Jr., business graduate and president of MIT; Edwin H. Land, president of Polaroid; and Jerome Wiesner, an engineer at the MIT Radiation Laboratory. According to the minutes of the meeting, recorded by an army general, Rabi informed the president of the suddenly dire competitive situation regarding the Soviet Union. Rabi stated in particular, "unless we take vigorous action they could pass us swiftly just as in a period of twenty or thirty years we caught up with Europe and left Western Europe far behind." Eisenhower, a former football player, immediately offered that he could present this to the American people as a kind of sports competition with the Soviets. According to the minutes, "he would like to try to create a spirit—an attitude toward science similar to that held toward various kinds of athletics."[1]

Eisenhower rallied his team in a televised speech to the nation on November 7. On the same day, at the urging of his advisors, he named James Killian to the new position of Presidential Science Advisor. Killian was afforded direct access to the president concerning scientific aspects of national security, space, and defense technology in particular. In addition, Eisenhower elevated the SAC to the President's Science Advisory Committee (PSAC), chaired by Killian. It was, in the words of one historian, the beginning of the "golden age of presidential scientific advising," a period that lasted only until the end of the Kennedy administration.[2]

Killian and his successors in that golden age came close to achieving the status and influence previously enjoyed by Vannevar Bush and similar scientist-administrators of the past, only now it was Killian and his full committee who exercised that authority. After the low point following the Oppenheimer case, scientists, especially physicists, were now nearly back on top.

The technological achievement that was *Sputnik* signified to many Americans the decline in the quality of the nation's science as well as of its technology. This appeared to put the nation in grave danger, both psychologically and militarily. Since, for many, technology meant simply applied science, a decline in science indicated a decline in the ability to keep up with the Soviets in the development of new weaponry and other technologies, despite the many billions poured into such work over the past decade. Now more than ever, prestigious triumphs in science, technology, and even in international sporting and cultural competitions were taken as evidence of a nation's political, economic, and cultural superiority. The Cold War had already become a competitive battle between ideological systems for prestige, power, and ultimate survival, especially in science and technology. As so often in the past, Americans responded to the national challenge with an enthusiasm that focused huge financial and human resources on the achievement of success at all costs.

MORE MONEY FOR MORE PHYSICISTS

During his speech on November 7, 1957, the president informed the nation that if the United States was going to compete head-on with the Soviet Union, it would have to give more attention "to science education and to the place of science in our national life."[3] The nation was appalled to learn that most American students did not take any physics at all in high school. Eisenhower called for the education of thousands more scientists as quickly as possible. In 1958 Congress passed the National Defense Education Act, providing $250 million for upgrading science education in public schools and for numerous scholarships for students studying in critical areas of the sciences and humanities, all in the name of defense. Students everywhere were urged to take physics and chemistry courses, and even those with less than stellar abilities were encouraged to enter careers in science, especially physics.

In 1960 Jerrold Zacharias at MIT and the Physical Science Study Committee, a panel of the nation's leading physicists, introduced one of the most innovative and influential transformations in the high school physics curriculum until that time. It was followed with great enthusiasm several years later by the Harvard Project Physics curriculum, a project developed by Gerald Holton and colleagues through extensive collaboration and classroom testing by educators and leading physicists nationwide. As the baby-boom generation began flooding into high schools, nearly every student who studied high-school physics encountered one or both of these curricula, and many did go on, or attempt to go on, to careers in physics. And many succeeded. The annual number of PhDs conferred in physics in the United States had declined from a high of 525 in 1954 to 450 in 1958. But beginning in 1959, as baby boomers and pre-baby-boomers responded to the call to science, the number of new physics PhDs experienced its biggest jump in history, reaching a record of 1,625 new PhDs in 1971. Unfortunately, only forty-eight of those doctorates went to women, and far fewer went to minorities. As indicated in Table 4 in the Appendix, even after passage in 1972 of the congressional bill mandating affirmative action, the numbers and percentages of doctorates awarded to women increased only modestly thereafter.[4]

Federal funding and with it jobs for the new doctorates kept pace with the increase through the first half of the 1960s. Shortly before the launch of *Sputnik*, a complacent Department of Defense (DoD) was preparing to reduce funding for basic research as the need for innovative new technologies seemed less urgent. Two days after *Sputnik*, Neil H. McElroy was appointed Secretary of Defense and immediately issued a directive making support of basic research once again a fundamental element of defense strategy.[5] In the following year, Congress and the administration established the National Aeronautics and Space Administration (NASA) to promote and administer non-military research and development related to space technology. In fiscal year 1962, the year Kennedy announced the goal of landing a man on the moon by the end of the decade, NASA commanded a budget of $1.7 billion, of which $196 million went to basic research.[6]

Federal funding for research and development in fiscal year 1957 (ending June 1957) had more than doubled since 1951 in inflation-adjusted dollars, owing in part to the impetus of the Korean War (see Table 2 in the Appendix). Of those funds, DoD-sponsored research still consumed the vast majority of the total, now about 74 percent, including the majority of funds funneled into physics research. *Sputnik* accelerated the pattern. Reflecting the new big-science initiatives, from fiscal year 1957 to 1962 the federal budget for research and development (including R&D plant), more than doubled in inflation-adjusted terms to $63 billion. It went up another 40 percent by fiscal year 1967 to the peak of the postwar years at over $88 billion, an amount representing nearly 11 percent of the entire federal budget for that year. Taking inflation into account, between 1959 and 1965 federal funding for basic research in physics tripled to $320 million. But as NASA consumed about a quarter of the research funding, the DoD portion of the total federal R&D budget dropped to about 47 percent after 1966 from its high of 74 percent during the 1950s, while the AEC budget, which included funds for the nuclear arsenal, hovered around 10 percent of total R&D funds.[7]

One result was the sudden expansion of most university physics departments. As new faculty members joined already existing research programs while others opened new lines of research, the collaborative American physics department became even less hierarchical and more democratic in nature. At the same time, the profession itself grew more fractious as disciplines and subdisciplines vied for recognition and resources. The prospect that one or a few powerful scientist-administrators could control the direction of the entire profession as they had in the past grew even more remote as the decade progressed.

As federal funds continued to flow into research, the domestic production of PhDs could not keep pace with the demand for new physicists.

Rather than tapping the still underrepresented segments of the American population, laboratories looked instead to foreign physicists, who began migrating to the United States in a so-called brain drain that foreign nations greeted with alarm. It was the beginning of the internationalization of the U.S. physics community that continued to the end of the century. About half of the jobs for physicists were at universities where research and the training of new physicists were in high demand. The remaining jobs were about equally divided between government and industry.

By the end of the decade, about one-third of the world's physicists worked in the United States, and they were publishing more papers in the leading physics research journal *Physical Review* than any other nation. The journal itself reflected the growth of physics in the United States and worldwide. Between 1945 and 1960 the *Physical Review* jumped in size from 300 pages per biweekly issue to 700 to 800 pages per issue. In 1964 the thick issues began appearing weekly. The first volume of *Physical Review Letters* in 1958 contained 500 pages. By 1993 it was 8,800 pages. In that same year, the *Physical Review*, which had by then split into different series by discipline, published over 70,000 pages of research![8]

U.S.-based physicists were also winning or sharing in more Nobel prizes in physics than any other nation. Many of them were awarded for discoveries made using the new accelerators. According to one count, between 1945 and 1987, fifty American-based researchers had received Nobel prizes in physics.[9] Thanks to the funding pumped into physics in response to *Sputnik*, American physics and physicists were once again at the top of the world—if they had ever really lost that position.

BIG MACHINES, BIG BUCKS, BIG THEORIES

Responding to "the call of the frontier," as historian Lillian Hoddeson and her coauthors put it, high-energy physics once again commanded substantial federal funds, even if the field still had little to do with new military or space technology.[10] Although the largest number of physicists, 25 percent, worked in condensed-matter research (formerly solid-state physics), and only about 10 percent worked in high-energy physics, over one-third of the federal funds for physics went to building and maintaining the prestigious high-energy accelerators.[11]

The familiar argument for funneling huge sums into high-energy research—that it would eventually trickle down to useful military and consumer technologies—now seemed to fade in comparison with the competitive spirit of the time and, for those in the field, compared with the exciting physics that the machines could produce. Like the space program, for some administrators the competition with the Soviet Union, and to a lesser extent with the new European accelerators (the Deutsches Elektronen-Synchrotron [DESY] near Hamburg, Germany, and CERN, the European Organization for Nuclear Research, near Geneva, Switzerland), became paramount after an attempt at international collaboration on a single world accelerator failed under mutual distrust.[12] In 1958 a panel of American scientists informed the government that Soviet scientists were now planning for an accelerator that could reach, not just the 10 billion electron volts that had concerned Frederick Seitz, but as much as 50 billion electron volts of energy, or 50 GeV (G for giga), more than double the biggest planned American accelerator at that time.

Killian's successor in the White House, George Kistiakowsky, helped convince the two main advisory committees, the president's PSAC and the AEC's GAC, that a 2-mile-long linear accelerator at Stanford University proposed by Wolfgang Panofsky was now a necessity. According to preliminary plans, the new machine would accelerate electrons and positrons (which would radiate away their energies in a cyclotron) to at least 25 GeV. After much debate, in 1959 Eisenhower asked Congress for $100 million in support of what became the Stanford Linear Accelerator (SLAC) National Laboratory. The new accelerators at university locations would continue to receive AEC rather than DoD funding; and they would act as hybrid laboratories, combining university research and oversight with national laboratory status.[13] When SLAC went operational in 1966, it was, however, far short of the 70 GeV of energy achieved a year later by a Russian proton synchrotron that began operation at Serpukhov, near Moscow, capturing the title of the world's most powerful accelerator.

Not to be outdone by the Soviet machine, the "arms race" in accelerators continued to accelerate as American high-energy physicists pressed even more urgently for the biggest machines ever. A proposal was already in development for a massive 200 to 300 GeV synchrotron to be built at a cost of $280 million, with an annual budget of $50 million to run it. This machine was needed, they argued, in order for the United States to maintain world leadership in science—no matter that a lot of world-class science was also being done outside of high-energy research. The 1963 official report of the joint PSAC/GAC committee recommending the new machine stated, "it is essential that the United States maintain its leading position in this area which ranks among our most prominent scientific undertakings."[14]

But critics, among them physicists Eugene Wigner, Alvin Weinberg, and Philip Abelson, complained that this one esoteric field, employing only about 10 percent of physicists, was draining resources from other research of equal interest and more practical value, such as condensed-matter physics, a field in which more physicists were engaged. Weinberg wrote in 1963:

> The field [high-energy physics] has no end of interesting things; it knows how to do them, and its people are the best. Yet I would be bold enough to argue that, at least by the criteria which I have set forth—relevance to the science in which it is embedded, relevance to human affairs, and relevance to technology—high-energy physics rates poorly. The world of sub-nuclear particles seems to be remote from the rest of the physical sciences . . . As for its direct or immediate bearing on human welfare, I believe it is essentially nil . . . high energy physics is expensive, not so much in money as in highly qualified people, especially those brilliant talents who could contribute so ably to other fields.[15]

Condensed-matter physicists argued that in their field of research the line between fundamental physics and its practical applications was so close that it was often blurred. With advances in laser, transistor, and semiconductor technology, a great deal of research was devoted to energy-band structure, quantum lattice vibrations (phonons) in crystals, and to a new technique: the implantation of ions into crystals using Van de Graaff accelerators, followed by studies of the nuclear magnetic resonance of the implanted ion as a tool for investigating crystal properties. But the publication of the BCS theory of superconductivity in 1957—named for its creators John Bardeen, Leon N. Cooper, and John R. Schrieffer—followed by the theoretical prediction of the Josephson effect in 1962, generated even greater interest and prospects for practical applications during the 1960s.

Superconductivity is the sudden decline of the electrical resistance in a wire to zero at an extremely low "critical" temperature. In this situation, a current passing through the wire would do so without any resistance, thus without any loss of energy. Brian David Josephson, a physics graduate student at Cambridge University, predicted that a direct current should be able to flow between two superconductors separated by a thin insulating barrier, even without an applied voltage. The current would cross the insulating gap by a process known as "tunneling," arising from the non-zero quantum probability that some current will get through. If a voltage is applied across the barrier, he predicted the appearance of an alternating current. The practical applications of superconductivity and Josephson tunneling were enormous. Confirmation of the predicted effects a year later by Philip Anderson and John Rowell at Bell Laboratories stimulated a flurry of research leading to a better understanding of the effect and its many potential applications in electronics and in the precision measurements of constants.[16]

Despite the promises of this research, Congress voted to funnel the greatest share of its physics funding to the proposed new accelerator. In 1965, after much debate between the East Coast, the West Coast, and eventually the Midwest, a site was selected in Weston (renamed Batavia), Illinois, near Chicago. In 1972 what became known as the Fermi National Accelerator Laboratory, or Fermilab, opened surprisingly on time and within budget. The big machine was soon accelerating protons to the world-record energy of 500 billion electron volts.[17]

But, of course, whatever the practical or prestige value of a big machine, a lot of good physics could be done with a powerful accelerator, and much of it did capture the public imagination (and still does). Rapid advances in high-energy physics accompanied each step in the evolution of the ever bigger machines. Despite its relatively small size, the field was then at the forefront of some of the most sophisticated experimental technology and some of the most elegant theories regarding such fundamental aspects of the physical world as the unity of the forces of nature, the symmetry of natural events, and even the origin of the universe.

Usually affiliated with large accelerator groups, particle theorists and experimentalists together were making strides toward a unification of the four fundamental forces of nature: gravitation, electromagnetism, the strong force that holds protons and neutrons together in the nucleus, and the weak force that controls the decay of individual neutrons. Each force could be represented by a field, and when each field was subjected to quantum mechanics, the resulting quanta were associated with elementary particles whose properties and combinations accounted for the properties of the forces generated by the fields. The ultimate theoretical goal was to merge the four forces together into a single force via relativistic quantum field theory, that is, a quantum theory of fields in which the requirements of relativity theory are satisfied. The unified force was assumed to emerge at extremely high energies, or temperatures, even perhaps at the temperature of the initial Big Bang, which was then still an unconfirmed hypothesis for the origin of the universe. At lower energies or temperatures, as in the universe today, these forces are separate.

In 1961 Sheldon Glashow at Harvard and Murray Gell-Mann at Caltech, drawing upon the earlier work of C. N. Yang, Robert Mills, and Julian Schwinger, discovered a way to combine the electromagnetic and weak interactions theoretically into what became the "electroweak" force. Six years later, Steven Weinberg at Harvard and Abdus Salam in Trieste independently developed a more elegant mathematical formalism for the electroweak force, in which the underlying symmetry (what is called the gauge symmetry) is spontaneously broken, yielding observable particles. They introduced a so-called Higgs mechanism into their equations, named for Peter Higgs of the University of Edinburgh, to account for the spontaneous breaking of the symmetry and for the masses of the elementary particles. The signature of the electroweak force would be the appearance of predicted new elementary particles in accelerator collisions at energies above 1 billion electron volts. Unfortunately, the new theory was so riddled with infinities that it seemed that not even renormalization would help.

At the same time as Glashow's and Gell-Mann's breakthrough of 1961, Gell-Mann and Yuval Ne'eman independently proposed a classification of all of the myriad observed elementary particles into groups of eight. A gap at one of the positions in these groups led Gell-Mann to predict the existence of a previously unobserved elementary particle. In 1964 a thirty-one person team of researchers at the Brookhaven National Accelerator Laboratory detected the predicted particle, and five years later, Gell-Mann received the Nobel Prize for physics.

Gell-Mann and, independently, George Zweig further theorized that the strong force could also be represented in a mathematically elegant fashion whereby all particles subject to the strong force were themselves composed of even more fundamental particles, what Gell-Mann called "quarks." They suggested initially that quarks would come in three "flavors": up, down, and strange. Three others eventually followed: the charmed quark, and the bottom and top quarks. Soon quarks were even endowed with three different "colors," making for numerous theoretical possibilities of combinations of quarks into different observed particles, especially for the strong force, and opening intensive research into the field of "quantum chromodynamics" beginning in the 1970s.

As new particle accelerators pushed protons and electrons to ever higher energies, several of the predicted particles began to appear. However, the theories predicting these particles did not get a firm footing until it was proved during the early 1970s that the renormalization procedure could in fact be extended to all of the quantum field theories then under consideration. Glashow recalled that it was not until the confirmation of the hypothesized charm quark through the simultaneous discovery of the so-called J/psi particle by research teams at SLAC and Brookhaven in 1974, that the earlier Weinberg-Salam-Glashow theory could be applied to the unification of the strong force with the unified electroweak force. The J/psi particle, unifying the names given the particle by the two teams, possessed a mass equivalent to 3.1 GeV of energy. It was identified as a bound charmed quark and charmed antiquark. This "grand unification" resulted in the so-called Standard Model, a theory that united three of the four forces of nature (except for gravitation) into a single theory.

While the up and down quarks were identified in earlier experiments at SLAC, Fermilab researchers identified the bottom quark in 1977 in order to account for the appearance of a new type of meson. In 1983 researchers at CERN, the European accelerator built to compete with the Soviet and American machines, reported the first possible evidence for the top quark, but because it is so massive (equivalent to about 174 GeV), confirmation had to await the development of Fermilab's trillion-volt "Tevatron" during the early 1990s. Upon receiving the Noble Prize for his work in 1979, along with Weinberg and Salam, Glashow remarked, "It's all one theory now. The strong, weak and electromagnetic interactions are all gauge theories . . . In a sense, it's all complete"—complete, of course, except for the gravitational force.[18]

Research has continued toward uniting the gravitational force to the other three forces, a task for which so-called string theory currently appears to hold some potential. A further task is to find the "Higgs boson," the supermassive particle that, according to the Standard Model, accounts, through the Higgs mechanism, for the existence of mass. It was for this purpose (and others) that, beginning at the close of the century, the biggest accelerator yet was built, the Large Hadron Collider completed at CERN in 2010.[19]

THE HONEYMOON IS OVER

President Eisenhower frequently referred to his science advisors as "my friends." After the low of 1954, a wider circle of his friends once again enjoyed influence in matters of national security, including defense programs, nuclear weapons policy, space technology, and even the defense partnership with physics. Moreover, because competition with the Soviet Union was also regarded as a life-or-death struggle between opposing political and economic systems, science advisors also enjoyed influence over social policies, including science education, allocation of resources, and federal research funding. This brought advisors back into the political arena, and inevitably back into conflict with other more powerful interests within government. This time, spurred by the events and upheavals of the 1960s, the public was more heavily involved. As during the early 1930s, pure scientists now faced increasing demands for research of social benefit, not just in service to military and commercial interests. Such demands arose not only from students, young scientists, and the general public, but even from government officials.[20]

John F. Kennedy had entered the presidency on good terms with scientists and the public. He appointed the first scientist to head the AEC, nuclear chemist Glenn T. Seaborg, and the scientific community warmly welcomed Kennedy's selection of Jerome B. Wiesner as his presidential science advisor. Wiesner also headed the new Office of Science and Technology Policy in the White House, an office that employed a large staff of scientists working on a host of technical policy issues for the administration. But, as earlier, trouble arose when Wiesner, a strong advocate for basic research and national defense, opposed Kennedy's plan to send a man to the moon. He argued that this huge, prestige-motivated undertaking would drain resources from much needed basic research and from more immediate social needs. He also advocated to Kennedy a slowing of the nuclear arms race through the negotiation of arms control and even nuclear disarmament.

At that time, the AEC's production of nuclear weapons was in high gear. The nation's nuclear arsenal climbed yearly to over 30,000 warheads by 1966 and reached the Cold War record of 31,255 warheads in fiscal year 1967, before dropping back below 30,000 in 1969.[21] (As of September 2009, the American nuclear weapons stockpile stood at 5,113 warheads.) Wiesner's advocacy brought him into conflict with the DoD and the National Security Council, setting off a new debate within the physics community over arms control and disarmament.[22]

Nevertheless, with help from Wiesner, Kennedy did manage to remove one of the public's most immediate concerns regarding nuclear weaponry, the dangerous radioactive fallout blanketing much of the nation from above-ground nuclear testing. In 1963, a year after the Cuban Missile Crisis had nearly brought the United States and the Soviet Union to war over Russian nuclear missiles stationed in Cuba, the two nations concluded their first nuclear treaty, the Limited Test Ban Treaty. Although limited, it prohibited nuclear testing above ground, in the oceans, and in outer space. Underground testing, which produced no fallout, could continue.

The near nuclear war over Cuban missiles, followed by the sudden easing of Cold War tensions through the limited test ban, enabled the public to breathe at last a cautious sigh of relief. As it did so, the public finally found its voice and began to question a nuclear policy that had brought the nation to the very brink of nuclear war. And it began to question the place of science, especially physics, within American society.

In the first real challenge to the tradition of "disinterested" pure science since the New Deal, the public began to demand the accountability of what some regarded as a secretive scientific elite that produced these horrible weapons of war while enjoying unprecedented research funding through federal tax dollars. Shouldn't research funded by society also benefit social needs? What is, or should be, the relationship of science to American democracy?[23] Scientists themselves began to question their limited role as producers of technical knowledge for use as others saw fit. Had military funding shifted the control of their work and their discipline to the Defense Department? Was it time to reassert responsibility for their work?

Congress and the public were appalled to learn in 1963 that 40 percent of all federal funds for academic research went to just ten universities, that only nine states located on the East Coast and the West Coast received 71 percent of government research funds, and that over half of the federal funds came from the DoD ($6.8 billion of the total research budget of $11.1 billion in 1962). To this, the director of the National Science Board, which oversaw the NSF, replied, "There is no point in putting money where there isn't competence to use it."[24]

To many, the contrast between the nation's needs and its priorities for science seemed completely out of balance. The nation's nuclear arsenal overflowed with weapons of mass destruction at the same time that industrial pollution, toxic waste, and uncured diseases were on the rise; substandard education, poverty, and discrimination still plagued some sectors of society; and the nation's technological infrastructure was in decay. Some began to blame science for the nation's ills, energizing an antiscientific revolt reminiscent of the early 1930s or the Romantic period at the beginning of the Industrial Revolution. Could "disinterested" pure science remain disinterested and pure when heavily funded by one government sector that expected returns on its investment? Shouldn't federally funded research begin to address the nation's other needs?

These questions had been around since the beginning of the century or earlier, but now they grew more intense as the war in Vietnam grew more heated. Ever since the Oppenheimer hearing, most scientists had been reluctant to express non–technically based convictions regarding government policy. This was now less the case. Matters came to a head after President Lyndon B. Johnson, with the approval of Congress, ordered a massive build-up of forces for the Vietnam War in 1965. As the unpopular war drew funding away from research and social programs, many of Johnson's science advisors felt obligated to speak out publicly against the war. An angered

Johnson responded by simply ignoring them and their advice. It was the beginning of the end of the "golden age" in science advising. When the imperial Johnson declared science to be too important and too expensive to be left to scientists, he began making science policy decisions himself, including the placement of new accelerators.

Politics began to enter explicitly into science. Because of this and the changing structure of the physics profession, the long-standing tradition of a few powerful scientist-administrators managing the course of physics and enjoying access to the highest levels of power came to an end. The nation's political leaders began to assume this role themselves, while large teams of big-science researchers exercised greater influence on research than did any individuals.[25] Some papers published in *Physical Review* already had close to 100 authors. As Spencer Weart noted, the small town of the physics community in which everyone knew each other had been replaced by the big city. But now, even the potential big-city mayors could no longer exercise the authority and influence that Bush, Millikan, Hale, and similar figures had once wielded over the nation's science. The physics community was now more democratic, but also, like most democracies, more chaotic.

In addition to the demise of the powerful science-administrator, the political and economic environments shifted rapidly during the second half of the 1960s as the escalation of the Vietnam War continued without pause, draining a federal budget further squeezed by Johnson's Great Society programs. The build-up also galvanized many young people, of whom men were still subject to the military draft; and it opened a fissure in public opinion, and among scientists, between those for and against the war. Beginning in 1968, when the federal budget deficit suddenly tripled to $25 billion, the financial crunch and the public criticism of science forced a constriction in federal funds for research that lasted for the next six years and led to the sudden bursting of the job bubble in physics and in other academic fields.[26]

The huge escalation in the number of physics PhDs following *Sputnik* had resulted by the late 1960s in such a glut of job seekers that, writes historian David Kaiser, "droves of new physics Ph.D.s slid into the worst job shortage the nation has ever seen," worse for physicists than during the Great Depression. During the annual meeting of the American Physical Society in 1971, 1,053 young physicists competed for just 53 job openings.[27]

As products of the post-*Sputnik* build-up in the number of physicists, many young scientists felt frustrated by their profession and restless and uncertain about their futures. While many left physics and science for other careers, others gravitated toward the student, antiwar, and science protest movements of the era.

As new technological weapons entered the unpopular war, scientists already opposed to further work on nuclear weapons found their views on the war and on the uses of their work at odds with the actions of their political leaders. Some joined political opposition groups emphasizing the social responsibility of science.[28] Late in 1968, fifty senior faculty members at MIT, the majority of whom were physicists and engineers, signed a declaration demanding a reassertion of scientists' responsibility for their work by redirecting its uses from weaponry to social needs. The following year they organized a "teach-in" at MIT to protest "the militarization of scientific research and promote science in the public interest." This led to the founding of the Union of Concerned Scientists, a lobbying and research group still active today.[29] Its first studies focused on defense-related research, nuclear weapons, and nuclear power. Later it took on energy, the environment, and the role of scientists in society.

For the concerned scientists, social engagement had completely replaced the elite self-reflection and academic disinterest of pure science. Others went even further. During the annual meeting of the American Physical Society (APS) in February 1969, three young faculty members and a postdoctoral researcher attempted, but failed, to induce the APS to issue a public declaration opposing the war in Vietnam. Over 300 of their colleagues joined with them to form what became Science for the People. To its founders, pure science was a pure fiction. They saw physics as one component of a nationwide "military-industrial-scientific complex" rooted in the fundamental economic and political relations of American society. These relations shaped the course of science and its uses in service to the agenda of the ruling complex of which it was a part, rather than in service to the needs of "the people."[30]

Although the APS refused to take a stand on the Vietnam War, debate within the APS had been mounting over the responsibility of the physicists' organization to society and over the role it should play, if any, in addressing controversial social issues. In response to a petition circulated in January 1969 by Brian Schwartz, then at MIT, the APS agreed in 1972 to establish a new division of the society, the Forum on Physics and Society. The forum provided at last an institutional vehicle for discussing social problems and concerns at meetings and for creating programs by which physicists could be address these issues. This resulted in the creation of committees on reactor safety and energy conservation, and related APS committees on professional concerns, minorities, education, and women in physics.

Schwartz also helped to initiate the last of these, the Committee on Women in Physics (now the Committee on the Status of Women in Physics). With a grant from the Sloan Foundation, its first order of business was a study revealing the clearly discriminatory practices against women employed in physics, a study that finally induced the APS Council to take a stand on a social issue. In February 1972 the Council issued a recommendation distributed to over 500 institutions employing women physicists urging "all physicists to press for equal acceptance, equal recognition, equal employment opportunity, equal advancement and equal salaries for physicists of equal ability and accomplishment without regard to sex."[31]

These developments among scientists, students, and the general public regarding the social responsibility of science paralleled and reflected a critical response within Congress to the role of federal funding in research. In 1969 Congress passed an amendment to the original law creating the National Science Foundation. It required the foundation to include support for the social sciences and for the social study of science as a means of rendering science of greater social benefit. Four years earlier, a House subcommittee had approved cutting 25 percent from the defense budget for basic research and redirecting the funds to social needs.[32] Also in 1969, Senator Mike Mansfield, a liberal opponent of the war who worried about the overdependence of academic science on military funding, pushed through an amendment to a military authorization bill that prohibited the military from financing any research not directly related to military objectives. Mansfield argued that, instead, the civilian National Science Foundation should fully support academic science. Nearly 20 percent of all academic research and 40 percent of all physics research was at that time funded by the military.[33]

Although the Mansfield Amendment was repealed a year later, it caused an outcry from many academic beneficiaries. But other academics and their students welcomed the removal of military influence derived from its long-standing funding arrangements with universities. As military funds decreased, and NSF funds for the social sciences increased, many universities divested themselves of at least a portion of their classified defense research. New academic disciplines, such as the sociology and the history of science, began to attain new professional status, and in the critical social environment of the day, historians and sociologists began to dismantle the apolitical, asocial, amoral ideology regarding the disinterested, value-free purity of physics. The utilization of social perspectives, historian Paul Forman argued at the time, was essential to achieving intellectual independence from physicists' constructs and practices. The historian's task required the association of those constructs and practices not to what he later called physicists' "transcendent" values, but to the physicists' social, institutional, and cultural environment. "The fiction of an autonomous development in the world of scientific ideas," wrote Forman, "has been maintained by systematically mistaking description for explanation, and by systematically refusing to look for or at contrary evidence." As one young protester put it at the time, "It is time to stop saying that science stands outside of society. Science is a social activity just like being a policeman, a factory worker or a politician."[34]

In response to the social criticism, the national weapons laboratories began to undertake socially useful research such as energy resources and environmental problems, and to advertise this shift to the public, a trend that continues to this day. In President Richard M. Nixon's first annual budget, some military research funds were replaced, not by unrestricted funds for pure science, but by civilian funds for practical civilian purposes. Lee DuBridge, an earlier advocate of pure science, summed up the monumental shift that had occurred: "The day is past when scientists and other scholars can sit quietly in their ivory towers unaware and unconcerned with the world outside their laboratories. Science is now a part of society, is a part of politics, is a part of the social and economic system."[35]

BANISHED AGAIN

Scientists greeted Nixon's election to the presidency in 1968 with mixed reactions. Most welcomed his appointment of Caltech president Lee DuBridge, a long-time Republican, as presidential science advisor and chair of the President's Science Advisory Committee (PSAC). But Nixon's continued escalation of the Vietnam War only inflamed tensions with scientists in and out of the White House. The tensions broke following Nixon's decisions to go ahead with the development of two controversial technologies, the Anti-Ballistic Missile (ABM) and the Supersonic Transport (SST). These episodes displayed the new confidence of scientists, mainly physicists, within the political arena, but also the dangers involved when undiplomatically confronting policymakers at the highest levels of government.[36]

It had been a goal of the military since the 1950s to develop a missile that could intercept and destroy an incoming nuclear ballistic missile. Because of the high speeds and brief time intervals involved, this type of missile required a quantum leap in tracking and guidance technology. Opponents, including those in the science protest groups, pointed out the destabilizing impact the development of an anti-ballistic missile would have on the arms race and on the "balance of terror" that had so far discouraged a nuclear war. DuBridge and the PSAC were already at odds with Nixon over the Vietnam War, and he and most of the PSAC, who were holdovers from Johnson's presidency, were already on record as opposed to the ABM when in February 1969 the DoD issued a report in favor of the ABM program. DuBridge and the PSAC were suddenly excluded from further advisory work regarding national security. But that did not stop them from going over the heads of the president and his staff in strongly opposing the ABM in public testimony before Congress. Nixon nevertheless gained congressional support for the ABM over the objections of his advisors and of most of the scientific community. Politically outmaneuvered, the scientists regarded the episode as evidence that the administration did not really want scientific advice at all, only the scientists' stamp of approval. For his part, Nixon saw the scientists as biased partisans pursuing their own agendas while pretending to serve as objective advisors.[37]

In the end, Nixon's strong-arm tactics did succeed in achieving a positive outcome. The United States and the Soviet Union signed the ABM Treaty in 1972, limiting most defensive missile systems. The new treaty became part of the SALT I agreement, for Strategic Arms Limitation Talks. (The United States withdrew from the ABM Treaty in 2002 in order to pursue a still unsuccessful anti-missile defense system.)

The Supersonic Transport proved the final straw. In addition to the prestige of landing a man on the moon by 1970, President Kennedy had set a goal by the end of the decade of building a commercially viable supersonic jet transport plane that was second to none. Likewise seeing the SST as a boost to business and national prestige, Nixon pushed for its development, again over the very public objections of his advisors and the very vocal objections of science protest groups. Aside from questioning the commercial viability, physicist William Shurcliffe was instrumental in raising objections centered on the damage caused by the jet's sonic booms, its impact on the ozone layer, and its huge expense without sufficient public benefits. Worried about the environment, Congress canceled the program. The United States never did develop such a transport.[38]

Fallout from the cancellation was inevitable. DuBridge, who had felt obligated to offer formal support of the SST, resigned in 1970. He was replaced by Edward E. David, a computer scientist from Bell Laboratories who had little political experience. Because the PSAC had overstepped its advisory role in publicly opposing announced government policies, Nixon was convinced the scientists were out to sabotage his administration. He placed some of its members on his secret "enemies list." At the end of his first term, Nixon accepted the formal resignations of the entire PSAC, and then failed to appoint any new members after he was reelected. In 1973 the science advisor also resigned, and his duties were returned to the director of the National Science Foundation, who had little direct contact with the president.

When Nixon disbanded the White House Office of Science and Technology Policy soon afterward, he effectively extinguished scientific advice in the White House. Congress, however, responded to these moves in the executive branch by establishing its own Office of Technology Assessment in 1972 for the purpose of providing Congress with sound scientific advice. (After Congress cut its funding, it ceased operations in 1995.)

In August 1974 Nixon himself resigned from office under the cloud of Watergate. Eight months later, American forces withdrew from Vietnam, ceding control of the nation to the Communist North Vietnamese. As with the Oppenheimer case, a chapter in American history and in American physics history had come to an end.

8
Revising the Partnership

The public critique of physics and the challenges of the Nixon years severely strained the partnership between physics and the federal government. As the economy struggled under debt and inflation, and as the nation struggled with the loss in Vietnam and the president's resignation, total federal funding for research and development (R&D) declined by 1974 to amounts adjusted for inflation not seen since 1962. Federal funds for education also decreased from 1968 through 1974, as did jobs for physicists. The annual number of physics PhDs conferred dropped by nearly half from the record high of 1,625 in 1971 to 862 in 1980.[1] The nation's international influence experienced a similar decline. A quarter century after the victory in World War II, the United States was in effect relinquishing its claims on the so-called American Century. The election of Jimmy Carter in 1976 signaled for physics the hopeful start of a new era.

Just months before Carter's election, Congress reestablished the position of Presidential Science Advisor and with it the Office of Science and Technology Policy, directed by the science advisor. But Congress chose not to revive the President's Science Advisory Committee, which had caused such controversy during Nixon's presidency. Nor did the science advisor play the same role as earlier. When Carter appointed Frank Press, a MIT professor of geophysics, to that position, Press understood what that new role would be. "From Frank Press on," writes Bruce L. R. Smith, "science advisors made totally clear that they served the president and did not represent the scientific community."[2]

A former naval submarine officer, Carter was the first president trained in nuclear engineering and the first since Kennedy with a commitment to fostering personal relationships with scientists. His secretary of defense was nuclear physicist Harold Brown, a former president of Caltech. Brown had grown up in New York City and received his doctorate in physics at Columbia University in 1949. He went on to positions at the Lawrence Berkeley Laboratory and at the Lawrence Livermore nuclear weapons laboratory, of which he was also director.

Under Carter, Press, and Brown, the administration pursued a program of technological modernization of the nation's forces, along with efforts to reduce Cold War tensions and the nuclear arms race. These led in 1979 to a second Strategic Arms Limitation Treaty, SALT II, negotiated between Carter and Soviet premier Leonid Brezhnev. But, shortly thereafter, the Soviet Union invaded Afghanistan. The treaty was never ratified by the Senate, although both nations abided by its provisions. In addition to the Soviet invasion of Afghanistan, the nation faced the U.S. hostage crisis in Iran, the humiliating failed rescue attempt by U.S. forces, an economic downturn at home, heightened economic competition from abroad, and the lingering memory of defeat in Vietnam. The public demanded that the nation exert a stronger military and economic presence on the world stage. Physicists, as the perceived source of new technologies, were once again in hot demand.

Ronald Reagan defeated Carter in the presidential election of 1980 with promises of renewing America's military and technological superiority and providing a more business-friendly economy. Federal funding for R&D once again rose rapidly, while federal taxes declined sharply. According to federal statistics, by 1989 inflation-adjusted federal R&D funding nearly matched the previous record amount achieved at the height of the Apollo space program in 1967 (see Table 2 in the Appendix). Those funds were divided almost evenly between basic and applied research. Defense Department funding increased even faster under Reagan, achieving 60 percent ($30 billion) of the total federal R&D budget in 1985, a percentage not seen since 1962 (after which the shift of DoD funds to NASA started to accelerate) and this did not include nuclear weapons research covered by the new Department of Energy (about 11.6 percent of federal R&D funding).[3] With so much new money flowing into research, the demand for physicists suddenly revived, and with it once again the production of new PhDs. By 1994 the annual production of physics PhDs (1,548) was nearly back to the record high in 1971 (1,625). But after that year it declined again as the partnership with government underwent yet another trying shift.[4] New physicists were on a roller-coaster ride during the last two decades of the century.

REORIENTING INDUSTRIAL RESEARCH

Beginning in the 1950s, approximately half of the federal funding for R&D went to national laboratories and academic institutions through grants and contracts. The other half went toward the funding of industrial R&D, primarily defense contract work. During the 1950s corporate funding of industrial R&D slightly outdid the federal government's total R&D support. The federal buildup of support following *Sputnik* naturally far surpassed industrial outlays. As federal funding declined after reaching its peak in 1967, industrial funding gradually caught up until, in 1981, corporate funding of industrial R&D surpassed the total federal R&D budget (see Table 5 in the Appendix). After that, there was no looking back. Between 1981 and 1989, inflation-adjusted industrial R&D funding jumped 52 percent, while federal funding increased by only about 33 percent. By 2000 corporations were providing 70 percent of the nation's total funding for R&D. What forces were driving this shift? What effects did they have on physics?

Most industrial research, especially within technology-based companies, had occurred within the semiautonomous central laboratories established earlier in the century—Bell Laboratories, General Electric (GE) Research Laboratory, International Business Machines (IBM) Research Division (founded in 1945), and others. As in national and academic laboratories, scientists were given wide latitude to follow their own interests, and for much the same reason: it was believed that independent pure research would inevitably lead to useful technologies. Still, big funding did not always lead to big results. Despite the higher salaries and no teaching load, corporations found it difficult to compete with government and academe for the best scientists, many of whom flocked to the prestigious fields of high-energy and nuclear physics. In addition, during the late 1960s and early 1970s some physicists harbored an aversion toward working for either business or government. A recent study undertaken by the Center for History of Physics (CHP) at the American Institute of Physics found that of the physicists who received physics doctorates in the period 1946 through 1965, nearly half (49 percent) held jobs in academe in 2001, while a little over a third (36 percent) were working in industry.[5] This pattern arose from circumstance as well as from choice.

The CHP study focused on physicists at fifteen major corporations employing roughly half of the physicists working in American industries. Extensive interviews, together with institutional and economic histories, revealed the profound transformation in industrial R&D that occurred during the latter decades of the twentieth century, as reflected in the huge increase in funding and in the employment pattern. The transformation entailed a shift in control of research beginning in the late 1960s from individual scientists working independently to department heads during the 1970s to the top management of the company during the 1980s and 1990s. This occurred as these companies fell under increasing competitive pressure that caused them to become more concerned with achieving financial returns on their investment in research rather than breakthrough discoveries in fundamental physics. As the emphasis shifted—much more explicitly than was the case with rising military funding—from free research to product development, both became more integrated with marketing and production. During the corporate takeovers and reorganization of the 1980s, followed by the "downsizing" of the 1990s, the semiautonomous central laboratories declined in importance. Many were replaced by product-oriented laboratories sponsored by individual corporate units, while many "pure" researchers transferred into administration or academe. The huge increase of industrial R&D funding since 1981 did not translate into commensurate breakthroughs in fundamental physics.

One of the prime examples of this process was the story of Bell Laboratories. As we saw earlier, Bell Laboratories was one of the most innovative and prestigious laboratories in the United States. Its six Nobel Prize winners had, among other things, discovered the transistor effect, contributed to the invention of the laser, and discovered the background radiation of the universe left over from the original Big Bang—hardly a discovery that would enhance telephone profits. When AT&T was declared a monopoly and broken up into seven regional telephone companies in 1984, it retained control over Bell Laboratories. But now under new competition, AT&T grew concerned that, according to the CHP report, 80 percent of the work at Bell Laboratories was unrelated to AT&T's primary telephone business.[6] Yet, ironically, when that research was telephone related, it worked against the company because the development of better products made telephone service cheaper. As the mobile phone business began its historic rise, AT&T decided in 1996 to spin off Bell Laboratories, along with most of its telephone manufacturing business, into a new company, Lucent Technologies. The work of the laboratories and their funding were now directly tied to corporate profits through product innovation in such areas as semiconductor research, high-speed networking, and electronics. Looking to the future economy, in 2010 the recently merged Alcatel-Lucent announced that Bell Laboratories would lead a consortium of telecommunications companies, chip makers, and university laboratories in the search for new technologies aimed at reducing the electricity consumption of communication networks by a factor of 1,000.[7]

When the computer and digital revolutions exploded onto the world stage beginning in the early 1980s, the parents of Bell Laboratories were not alone in shifting resources from free-form pure research directly to the development of product-oriented technologies. In 1977 the recently founded Apple Inc. introduced its first successful micro or personal computer, the Apple II. Its commercial success was tied to its user-friendly operating system and to the VisiCalc spreadsheet program designed for the Apple II, making it attractive to business users. In 1981 IBM, until then a producer of big main-frame computers, introduced the competing IBM Personal Computer, or PC, in partnership with the start-up software company Microsoft. The fledgling company had purchased and enhanced a disk operating system (DOS) for the new machines, calling its version

MS-DOS. In 1985 the laser compact disc, or CD, was first introduced. In 1988 the federal government opened the Internet to public commercial use, and in 1991 Tim Berners-Lee, a physicist CERN, the European Organization for Nuclear Research, introduced the World Wide Web as a publicly accessible service on the Internet.

Beginning in the mid-1990s, when Congress encouraged competition between telephone and cable companies, these companies began to replace the copper wires in the backbone of the Internet with high-speed, high-bandwidth, fiber-optic cable capable of handling the rapid transfer of huge amounts of digital data. As consumers rushed to buy the latest high-tech devices, numerous old and new technology companies promoting innovative computer and Internet-related products generated an enormous sense of excitement and creative vigor, yielding new digital consumer products ranging from computers and networking devices to mobile phones, iPods, and iPads. Running on the digital light pulses of fiber-optic cable, the Internet was now truly global.[8] So, too, was the instant global interchange of new data and research results, contributing to the increasing globalization of physics and all sciences during the latter decades of the century. Telescopes and even accelerators could now be controlled remotely, and their images and data could be downloaded and analyzed "in real time." Researchers in any country with an Internet connection could now gain immediate access to online data, and papers could be posted online for immediate review and critique.

Because transistors, integrated circuits, and high-speed switching devices lay at the foundation of the new technologies, a burst of research in the physics of these and related devices occurred. It paralleled the discovery of high-temperature superconductivity at an IBM research laboratory in Zurich, Switzerland, in 1986. Superconductivity had been observed only at extremely low temperatures, close to absolute zero. The prospect that superconductivity in some materials could occur at somewhat higher temperatures, perhaps even close to room temperature, raised the prospect of commercially viable, super-fast, low-energy computers and networks. It also meant the prospect of higher temperature superconducting electromagnets whose huge magnetic fields would enable the movement of high-speed trains on tracks without any friction, so-called maglev trains. Such magnets would also bring down the cost and difficulty of maintaining high-speed elementary particles on the circular paths of cyclotron accelerators.

Suddenly, in those decades, condensed-matter physics and the related field of materials research became areas of choice for many of the new PhD physicists as new jobs opened and money flowed within the high-tech industries. While many of the earlier physicists had entered and remained in careers in academe, of those who earned physics doctorates in the years 1996 through 2000 the situation was reversed: more than half (57 percent) were working in industry in 2001, while less than a third (31 percent) were in academe. The same held for the employment distribution of the profession as a whole. Of the estimated 37,000 physicists in the United States in 2001, approximately one-third were employed in academe, while the rest were employed elsewhere, mainly in industry as well as in government, nonprofits, and other organizations.[9] In 2000, of the 1,205 doctorates conferred in physics by U.S. universities, 41 percent were in condensed matter, materials science, and related fields, and only 30 percent were in the more academic and government-sponsored fields of astronomy/astrophysics and particles/fields. By 2005 the most popular units of the American Physical Society were by far the Division of Condensed Matter Physics and the Forum on Industrial and Applied Physics.[10]

COMPUTING PHYSICS

During the 1980s, the last decade of the Cold War, the war grew even colder as tensions increased. Still, hints of the coming thaw were emerging as some branches of physics participated in the new tendency toward multidisciplinary and even multinational research beyond the scope or perspective of any one discipline or nation. At the same time, the composition of the physics community in the United States became more international.

Two of the prominently researched phenomena in this period, which also held considerable global significance, concerned the environment: the problem of global warming and the human contribution to it, and the prospect of global devastation arising from a "nuclear winter" triggered by a nuclear war.[11] Both phenomena arose from human activity and both entailed profound political, economic, and social consequences. But because these were such complex, large-scale, and long-term phenomena with many different measured variables, precise predictions of their effects required large-scale computer modeling and the constant fine-tuning of the assumptions and variables used in the models and simulations. The development of new "supercomputers" during the previous decade made these calculations possible.

Since the depths of World War II, physicists had used digital computers for decryption, ballistics, data analysis, and calculations related to the construction of nuclear weapons. After the war, the Defense Department remained almost the exclusive sponsor of computer development, funding the construction of large-scale machines designed primarily for the computation of controlled and uncontrolled nuclear fission and fusion reactions. The early big machines became more adaptable to these purposes when they utilized the so-called von Neumann architecture, named for physicist and mathematician John von Neumann. In this arrangement, the data and the program to handle them are both stored in the computer's memory, rather than stored separately, thus vastly speeding up the analysis process. Military benefactors in partnership with industry provided the primary impetus and funding for ever bigger and faster computers using this architecture, often at Atomic Energy Commission laboratories. Physicists were involved in all aspects of their development and use.[12]

In 1952, von Neumann, who had earlier worked in hydrodynamics, built an advanced computer at the Institute for Advanced Study in Princeton that was used, in addition to military research, for modeling hydrodynamic flow within the atmosphere in connection with meteorological forecasting. This work stimulated a quantum leap in numerical meteorology and an impetus to the forecasting skills of the National Weather Bureau (now Service).[13]

Meanwhile, at accelerator laboratories, bubble chambers had become the detector of choice for capturing collision events. Luis Alvarez, head of the bubble chamber group at the Lawrence Radiation Laboratory at Berkeley, pioneered methods for data extraction and analysis from the thousands of bubble-chamber photographs generated by a single experiment.

As the capability of computers advanced, by the mid-1960s computers had become essential for data acquisition and analysis in particle physics and in other data-intensive fields.[14] Peter Galison has shown that the effect of the widespread utilization of computers in high-energy experimental research both for acquisition of data and for its analysis was to eliminate the experimenter from the experiment.[15] While computers enabled faster and more reliable acquisition of ever larger blocks of data, the discipline reflected to an even greater degree the production factory model of large-scale, big-machine physics.[16]

The heavy "number crunching" involved in complex modeling and in data acquisition and analysis required the development of dedicated computers. With generous military funding, IBM led the way into the realm of these so-called supercomputers. The company began small in 1956 when it released the IBM 704, a computer with a magnetic core memory but no operating system and only a rudimentary assembly language compiler for programming the machine. A year later, the company released a primitive operating system for its computer, along with a compiler for its newly invented high-level FORTRAN programming language (for *For*mula *Trans*lating System), making it perhaps the first user-friendly device for routine calculations. In 1960, the IBM 7090 incorporated solid-state electronics for the first time; in 1965, IBM released the IBM 360 mainframe with local time-sharing terminals that made it the mainstay, as the company's name implied, of routine business as well as scientific computing needs. Although IBM also built bigger machines for the military as part of its STRETCH series during the early 1960s, the 360 became its main development and marketing focus.

This left other companies to fill the gap in the production of ever faster machines handling the intensive computations required for physics-related applications. In 1964, the Control Data Corporation (CDC), founded in 1957, introduced the CDC 6600 for this purpose. Designed by Seymour Cray, it was regarded at the time as the first widely used supercomputer. It could perform over 3 million floating point operations per second (Mflops). It achieved this by the introduction of "functional parallelism," the splitting of an application into separate parallel but interrelated processing units. Cray left CDC in the early 1970s to found his own company, Cray Research, which produced in 1976 the Cray-1 computer that reached 250 Mflops.[17]

As supercomputers became available beginning in the mid-1960s, physicists and mathematicians became their primary users, and not only in military laboratories. Some of the users became interested in the computers themselves, providing a huge impetus to the study of computer science. As with the transistor and integrated circuits, physicists became involved in nearly every aspect of hardware development. As computers entered physics research, computational physics became an indispensable tool of the physicist. This was so much the case that, in some fields, computer calculations and simulations joined theoretical and experimental research as a third mode of scientific inquiry.[18]

By the 1980s, with the advent of accelerators that used colliding beams of particles rather than a single beam projected onto a fixed target, data processing, storage, and analysis were demanding innovative new strategies. In high-energy physics, solid-state detectors were replacing bubble and spark chambers, and accelerators were producing as many as a billion collisions per second. Even more powerful supercomputers were essential for gathering the data and rapidly analyzing the tracks of newly created particles. In one type of detector, particles newly created by the colliding beams pass through layers upon layers of tiny silicon diodes. Much as light is captured by the "pixels" in the sensor of a digital camera, each diode emits a tiny current whenever a particle passes through it. Together, these tiny currents enable the computer to reconstruct the path of the particle from which the particle's charge, momentum, and possibly even its identity can be determined.

However, one practical difficulty with this scenario is that the colliding beams, such as those produced by the Tevatron at Fermi National Accelerator Laboratory (Fermilab) in Illinois or the Large Hadron Collider at CERN, near Geneva, Switzerland, produce so much data for storage and analysis that no single computer, however "super," can handle it all at a manageable cost. Instead, during the 1980s, with the advent of smaller "personal computers," Robert Seidel has argued, a new division of labor was introduced. Supercomputers gathered the data, but "farms" of small computers operated by individual researchers or small teams of researchers were employed to analyze smaller chunks of the data. With the advent of the Internet, "distributed" computing became the norm, bringing physicists back into control of their experiments and the computer analysis of their data, and at a lower financial cost.[19]

The Large Hadron Collider at CERN utilizes what its managers call "grid computing." Even before the machine was completed, they established a worldwide network of computers at 130 sites in thirty-four countries to work together through the Internet to process, store, and analyze the data almost as quickly as it is produced by the big machine. This "massive distributed computing infrastructure" provides over 8,000 physicists around the world with nearly real-time access to the accelerator data. Such a model most likely represents the future of supercomputing applied to data-intensive experiments, and, at the same time, it represents a further instance of the continuing globalization of fundamental research.

COMPUTING NUCLEAR DISASTER

In 1982, the NASA Ames Research Center at Moffett Field, California, purchased a new Cray X-MP computer (at 400 MFlops, the world's leader until 1985). During this period at Ames, writes the center's historian, "supercomputing permeated everything so that computer codes seemed to replace the scientific theory that had earlier guided so much of what Ames did."[20] It was on this machine that in 1983 a team of physicists and climatologists at Ames reached a startling conclusion.

Several years earlier, Ames scientists Owen Brian Toon, Thomas Ackerman, and James Pollack, together with consultants Richard Turco at R&D Associates and Carl Sagan, an astrophysicist at Cornell, had been modeling the atmosphere of Mars using an earlier Cray supercomputer and data provided by the Mariner 9 mission to Mars. The team—often referred to together as TTAPS, after the first letters of their last names—was drawn closer to home when they undertook a computer study of the climatic effects of the 1980 Mount Saint Helens volcanic eruption. They also undertook a computer simulation of the asteroid hypothesis for the extinction of the dinosaurs that had been put forth, also in 1980, by physicist Luis Alvarez and his son Walter, a geophysicist. When the new Reagan administration began speaking of a nuclear war as "winnable," or at least "survivable," the Ames researchers began investigating these assertions by adapting their models to the atmospheric effects of a nuclear war. As Lawrence Badash has described, they utilized a combination of three computer-based models: a nuclear-war scenario; a particle microphysics model of the dust and sooty smoke of burning cities produced by a nuclear war; and a radiative-convective model of the atmosphere that was, however, as yet only one dimensional (for the vertical dimension). Combining the lessons of their previous studies with physical meteorology, atmospheric chemistry, and studies of forest and urban fires, their computer models utilizing the new Cray X-MP yielded the startling result that, Sagan informed a press conference in Washington, D.C., on Halloween 1983, a nuclear war could possibly trigger what he and his colleagues were calling a "nuclear winter."[21]

Depending on the number and size of the nuclear devices exploded over urban centers in a nuclear war, TTAPS predicted that nuclear fireballs would carry millions of tons of dust and sooty smoke high into the atmosphere, darkening the skies to a few percent of normal sunlight and nearly depleting the ozone layer. Ground temperatures over the globe would drop to −15° to −25°C and remain there for months. In an accompanying report, environmentalist Paul Ehrlich and colleagues (including Sagan) predicted a collapse of the world's food supply owing to the cessation of photosynthesis. Even after the atmosphere cleared, all life on the planet would be swept by deadly ultraviolet radiation, high radioactivity, and toxic chemicals. Many species would go extinct, including, quite possibly, humans. While an estimated half a billion people would die in a nuclear war, billions more would die in the aftermath. Far from being winnable or even survivable, nuclear war was most likely tantamount to nuclear suicide.[22]

Such horrific predictions came at a time of increased anxiety regarding nuclear war and nuclear physicists. As others took up nuclear-winter research in the United States, the Soviet Union, and on the U.N. Scientific Committee on Problems of the Environment, Sagan left his coworkers behind to become practically a one-man public relations promoter of the nuclear winter phenomenon and of the need to prevent nuclear war at all costs. Physics and politics had converged. Even as his arguments went beyond the scope of the study, however, they still exerted little impact on the direction of the nation's nuclear policies.[23]

What made the nation so anxious were not only the rising Cold War tensions resulting from the president's mandate to bring the nation to greater military stature, but also two of the president's most controversial decisions. Both of these occurred in the same year as the nuclear winter report, and together they brought physicists once again into the arena of public debate. In 1983, the United States began to station middle-range

nuclear rockets in Western Europe that were capable of reaching the Soviet Union. This set off protests in Europe and the United States and galvanized the nascent "nuclear-freeze" movement. In March, President Reagan announced that the United States was launching a multi-billion-dollar
effort to develop a defensive nuclear weapons system called the Strategic
Defense Initiative (SDI), popularly known as "star wars." The search for a
suitable missile defense system had nearly halted in the early 1970s after
the Anti-Ballistic Missile Treaty of the Nixon era. But such a system was
still unworkable in any event. An incoming enemy missile could now deploy as many as ten warheads, some of which may be dummies, but all of
which had to be destroyed for an effective defense.

The technical feasibility of a missile defense system suddenly gained
new traction when nuclear physicist Edward Teller reported to Congress
and the president in 1982 that scientists at the Lawrence Livermore National Laboratory had produced the first X-ray laser beam capable of
destroying a warhead.[24] Such a device required so much power that it
utilized the directed energy of a small hydrogen bomb. Teller and several
colleagues lobbied Congress and the president for a new Manhattan Project to build a laser defense system. Congress was skeptical of the idea because it would violate the ABM Treaty, thereby further increasing tensions with the Soviet Union. In addition, Reagan's own science advisor,
George A. Keyworth, a nuclear physicist from the Los Alamos National
Laboratory whom Teller had recommended, was himself skeptical of the
plan and of Teller's claim that the new weapon would be available within
five years.

After Teller assured the president that the system was indeed feasible
and that it would be ready in the foreseeable future, Reagan announced in
1983 a crash program to build it. The United States subsequently withdrew from the still unratified START II treaty. Reagan's announcement
set off a storm of protest, not only among physicists, but also from the
general public in the United States and abroad. In addition to the destabilizing effect of SDI, most physicists were concerned that Teller and sup-
porters had oversold the idea to government leaders. Such a system would
require the detection of a missile within minutes of its launch, tracking
the missile as it reentered Earth's atmosphere from space orbit at nearly
20,000 miles per hour, a high-speed supercomputer to track all of the
multiple warheads emitted, and the firing of bursts of intense radiation at

each within seconds. Such a sophisticated system seemed far beyond the current technical capabilities of lasers, supercomputers, and even tactical practicality. In 1987, a distinguished study group of the American Physical Society, now more engaged in social issues, concluded: "even in the best of circumstances, a decade or more of intensive research would be required to provide the technical knowledge needed for an informed decision about the potential effectiveness and survivability of directed energy weapon systems."[25] Nevertheless, the nation invested an estimated $3.5 billion in SDI over the next three years, creating numerous research jobs and helping to spur the laser and computer technologies that are now central to the information age. But, with success nowhere in sight, the project eventually lost funding, although the idea never quite died. It was revived again during the Bush administration of the early 2000s.

By the same token, the idea of a nuclear winter slowly faded during the years ahead, though it never quite died. One of the problems was the large uncertainty in the parameters used for the computer models. A lot depended, of course, on how many nuclear weapons would be released in a war, on what types of targets, what season of the year, and so on. As with global warming, the imprecision of the parameters and the long-range nature of the predictions left enough room for some to doubt the results. In 1990, as the Cold War neared its conclusion, the authors of the original study used a three-dimensional model of the atmosphere for the first time and discovered that the nuclear winter effect would actually be somewhat milder than originally believed, more like a "nuclear autumn."[26] More recently, Toon, Turco, and other colleagues have shown that a local or regional nuclear war, such as one in the Middle East, would have the capacity to set off at least a nuclear autumn. Although less severe than a nuclear winter, it would still cause the starvation of a sizable fraction of humanity, resulting, they proclaimed, in the equation "local nuclear war = global suffering."[27]

THE RISE AND FALL OF THE SUPERCONDUCTING SUPER COLLIDER

The end of the Cold War exerted a profound and far-reaching effect on high-energy physics and, as earlier, on the overall partnership of physics with the federal government.

Since the end of World War II, high-energy physics, despite representing only about 10 percent of the physics community, had enjoyed a special relationship with the federal government. Fundamental physics had helped win the last world war, and particle physics was argued to be one of the most fundamental of all branches of fundamental physics. But particle physics required the building of huge, expensive accelerators that only the federal government could afford. Entering into its postwar partnership with high-energy physicists, the federal government had taken on the task of supporting ever bigger accelerators in the expectation that, as in the world war, future discoveries might somehow greatly benefit national security. In addition, the world's biggest accelerators and the Nobel prizes they generated brought international prestige and demonstrated the superiority of American science and culture over those of its Cold War competitor, the Soviet Union. Such ideals once again motivated federal support during the 1980s for the biggest and most expensive machine to date, the Superconducting Super Collider, or SSC. But the end of the Cold War and the breakup of the Soviet Union in 1991, together with other economic, political, and scientific factors, forced the demise of the SSC in 1993, altering forever the postwar partnership between physics and the federal government.

In physics research, theory and experiment often go hand in hand. The aim of particle physics is to comprehend the structure of individual atoms, nuclei, and their constituents at the most fundamental level of the particle building blocks of matter. These building blocks are governed by the fundamental forces of nature: gravitation, electromagnetism, and the strong and weak nuclear forces. By the early 1970s, the last three of these forces had been joined together theoretically into an overall theory known as the Standard Model. According to this theory, as noted earlier, many so-called elementary particles, such as the protons and neutrons that make up nuclei, are in fact composed of the more elementary particles quarks, gluons, and leptons. In addition, a special particle called the Higgs boson accounts for the existence of mass itself. Such fundamental particles can be observed only when everyday protons, neutrons, and mesons collide together at extremely high energies. Einstein had shown that mass and energy can be converted into each other. Because of this, when two protons collide at high enough energy, their masses melt together and then condense into the particles predicted by the theory. Such extreme conditions had not existed since the Big Bang at the origin of the universe when, it was also suggested, the four forces of nature were initially united. As the universe cooled, the forces separated, giving rise to the different particles and aggregations of particles that make up the matter that we observe today in the universe. Such ideas captured the imagination of the public as well as that of the scientists, helping to encourage the expenditure of billions of tax dollars for ever bigger accelerators.

By the late 1970s, most of the predicted quarks and leptons had been observed in accelerator experiments, but some were still missing. More-over, the Higgs particle that accounts for mass had yet to be observed, nor had three other particles been observed, also known as bosons, that represent the unity of the electromagnetic and weak forces. In an attempt to find these predicted particles and to maintain the nation's competitive leadership in this prestige field, federal funds flowed into American accel-crator facilities during the late 1970s. While Fermilab managed to double its maximum energy to 1 trillion electron volts and the Stanford Linear Accelerator (SLAC) pushed to higher energies, the Department of Energy funded construction of a new accelerator at Brookhaven National Labora-tory called Isabelle, endowed with superconducting electromagnets that would enable it to achieve several trillion electron volts.[28]

But the United States was not alone in its quest. In Switzerland, CERN was already achieving comparable energies just as difficulties with Isabelle's magnets brought delays. During a conference in 1982, Fermilab director Leon Lederman proposed a new, internationally funded superconducting-magnet accelerator, the SSC, to be built in the United States to attain even higher energies. As this proposal circulated in 1983, the CERN accelerator stunned American physicists with the detection of the three long-sought bosons. Suddenly, the United States had fallen behind in the race for particle-physics prestige.[29]

The competitive and scientific environments once again required that the United States regain scientific preeminence. President Reagan's sci-ence advisor George Keyworth seized on the proposed SSC as the means to do just that. American particle physicists echoed the sentiment, which only reinforced their eagerness for the new physics the machine would produce. In 1985, Glashow and Lederman wrote in *Physics Today*, "Of course, as scientists, we must rejoice in the brilliant achievements of our colleagues overseas. Our concern is that if we forego the opportunity that the SSC offers for the 1990s, the loss will not only be to our science but also to the broader issue of national pride and technological self-confidence. When we were children, America did most things best. So it should again."[30]

Planning for the SSC began in earnest in 1984 after the Department of Energy terminated Isabelle and transferred its funds to planning for the SSC.[31] Two years later, the design group at the Lawrence Berkeley Laboratory submitted its report for a machine to be built by the early 1990s at a cost of $3 billion. The huge underground device would be 52 miles in circumference with two detectors larger than battleships. It would employ thousands of workers of all types, not just scientists. Protons would accelerate around two rings in opposite directions to energies of 20 TeV before colliding into each other within the detectors. This was sixty times more than the energies then achieved at CERN. The ten-year construction project would restore the nation to preeminence in high-energy physics and enable the study of the physical world at the most fundamental level. It would also, it was argued, produce a new generation of highly trained physicists, foster the continued growth of computer technology, and contribute to the development of new medical applications.

In 1987, President Reagan personally approved the SSC, which now had a budget of $4.4 billion. By the time construction started on a site selected in Texas, the cost had jumped to $5.9 billion. In 1991, with a slightly larger size, it was $8.25 billion. Congress still supported the effort, but it demanded a sizable contribution to its cost from nonfederal sources, including foreign nations. By 1993, after President Clinton proposed a three-year extension of construction to reduce annual costs, the total cost was projected at nearly $11 billion—half the size of the Manhattan Project, adjusted for inflation. But by then political support had collapsed. On October 19, 1993, the House voted to kill the project after the expenditure of several billion dollars on construction. Contracts were canceled, jobs ended, and the partially completed tunnels under the Texas prairie were left empty.

Many factors contributed to the sudden death of the SSC. The most obvious was the rapid growth in the cost accompanied by the failure to obtain any funding from foreign nations as required by Congress. Seeing the SSC as a symbol of national pride and power, the administration had made little effort to encourage foreign funding through diplomacy or by outsourcing work to foreign companies. In addition, the federal budget deficit had ballooned as the Reagan administration undertook massive spending for defense and defense-related R&D while reducing taxes without reducing other expenditures significantly. As early as 1982, the annual budget deficit exceeded $100 billion for the first time in history, and it continued to climb to nearly $300 billion by 1992, setting off a recession.[32] The SSC was an easy target for budget-conscious House members, one of whom doubted "that the most pressing issues facing the Nation include an insufficient understanding of the origins of the universe, a deteriorating standard of living for high-energy physicists, or declining American competitiveness in the race to find elusive particles."[33]

In response to the looming deficit, Congress passed a law in 1990 that imposed caps on defense and discretionary spending, including science. The Energy Department cut its high-energy physics budget by 10 percent. This meant that additional funds flowing into the SSC had to flow from other research. Once again resentment among other physicists was building against high-energy physicists for their seemingly excessive power and money. Arguments reminiscent of those raised in the 1960s abounded. Some of the critics were distinguished condensed-matter physicists such as Nobel Prize winner John Schrieffer, the S in the BCS theory of superconductivity. Schrieffer argued that condensed-matter physics was just as fundamental for aggregate matter as was particle physics for particles. Philip Anderson, another Nobel Prize winner, testified before Congress, "dollar for dollar we in condensed-matter physics have spun off a lot more billions than the particle physicists . . . and we can honestly promise to continue to do so."[34]

Anderson had hit upon perhaps the most significant reason for the failure of the SSC. Condensed-matter physicists could promise the likelihood of eventual practical consumer and defense products from their research as a return on the investment of federal and corporate funds. But high-energy physicists could no longer point to the likelihood of significant contributions to national security or to the economy as a reason to fund their expensive machines.

The Berlin Wall fell suddenly in 1989. By 1991 the Cold War was over. The United States was no longer engaged in scientific, cultural, or military competition with the Soviet Union, and Congress was already re-examining the place of science in the post–Cold War landscape. Daniel Kevles has argued that the vote against the SSC was not directed against science or against the long-term partnership of science with the federal government, "rather, it signified a redirection of the partnership's aims in line with the felt needs of the post–Cold War circumstances." Curiosity-driven "pure science" would continue, but Henry Rowland would not have been pleased to learn that pure science could no longer remain, or claim to remain, completely insulated from the political, economic, and social needs and demands of its society. The emphasis now shifted to "strategic" or "targeted" areas of research, writes Kevles, "fields likely to produce results for practical purposes such as strengthening the nation's economic competitiveness or its ability to deal with global environmental change."[35] Having once occupied a privileged place in federal planning, American physics and science would have to compete for limited resources just like any other interest group within American society.

CRITIQUING CONTEMPORARY PHYSICS

The demise of the SSC and the gradual retreat of physics and other "pure" sciences from their exalted, socially and politically insulated positions during the last third of the century paralleled the rise of the history of science profession to a more mature status, a profession that had become more independent and more intellectually and socially critical of its subject. Perhaps reflecting the state of affairs, one author, a science journalist, went so far as to proclaim in a popular book "the end of science" and "the limits of knowledge in the twilight of the scientific age."[36] At the same time, the earlier sociological and institutional approaches to physics history had gradually transformed into the "science studies" approach arising in the late 1980s, whereby, very briefly put, the research results produced by a community were seen not as universal transcendent truths, but as social constructs contingent upon the sociological structure and aims of the community.

From an entirely different direction, the nature and intellectual import of physics research also came under closer scrutiny through philosophically based interpretative analyses of the intellectual structure of the research enterprise and its results. Perhaps because of the attention given the fateful SSC and the prima donna status of its founders, both perspectives tended to focus at first almost exclusively on high-energy particle physics. Both pointed, however, toward the emergence, as in physics earlier, of two new hybrid disciplines. One of the disciplines combined traditional history of science with the sociology of science in the "science studies" movement; the other joined intellectual history, cultural studies, and the philosophy of science into what might be called "analytic history of science."[37]

In line with the analytical and sociological approaches, Paul Forman began to pursue in that period an influential methodological critique that contrasted the work and aims of physicists with the necessary independent scholarship of the historian of physics. Drawing upon his earlier critique of the ideological commitments of the physics discipline both in the United States and in Weimar Germany, Forman offered a broader analysis of the aims and outlook of the contemporary American physics discipline. It was enabled, he wrote, by a perceived cultural shift beginning in the early 1980s from the modern era to the so-called postmodern era.[38] This shift revealed to him more clearly that scientists' insistence on the value-free nature of their work and their avoidance of moral responsibility for its results were, in fact, "an important part of the project of modernity."[39]

We have already observed in earlier chapters how nuclear physicists were, willingly or not, stripped of their moral responsibilities during World War II and prevented from exercising those responsibilities in the use of the atomic bombs and in the development of the hydrogen bomb. Such treatment could be seen as having prepared these and other physicists for alignment with the project of value-free modernity identified by Forman. The alignment was reinforced, he argued, by the modernist social agenda of the postwar era, encouraged, one might add, by the virtually unlimited federal funding of research. Apparently even the reawakening of some physicists to their social responsibilities at various times during the postwar era did not suffice to break the hold of value-free "transcendence."

In creating the ideology of pure science, we have seen how scientists since Plato have frequently exploited the ideals of objective, value-free scientific truth transcending the mundane, politically messy, and morally demanding world of everyday existence in order to argue for the lofty status and independence of their discipline. Forman suggested that by adhering to the transcendence of scientific truths produced by the physics discipline of which one is a member, the physicist ultimately surrenders his or her moral independence, investing moral authority in the discipline itself. Because scientific truth is held to transcend everything else, the physicist can believe himself or herself relieved of all responsibility, even as he or she becomes entangled, along with the discipline, in work of obvious political, moral, and social significance. This characteristic of the modern era, Forman argued further, is the reason that many scientists, and even many historians of science, have until recently failed to recognize, or accept, that the "siting of knowledge production has significance for the character of the knowledge produced." In other words, as he argued with respect to postwar electronics research, the sources of funding and the sponsorship of research inevitably affect the direction and content of the results produced: "in short, power creates knowledge in its own interest and its own image." Despite the spurt of critical sociological thinking about science during the 1960s and more recently, only with the onset of the postmodern era are historians and even many physicists finally "coming to accept that 'pure' knowledge is an illusion (because purity is eo ipso transcendence)."[40]

Thus, in addition to occupying a less exalted social and political status during the last decade of the century, physics and the knowledge it produces had likewise come to occupy a less transcendent intellectual status as one among many other intellectual disciplines. Like them, it was now more often regarded by those who study the work of physicists as subject to the same social, ideological, cultural, economic, and anthropological forces and agendas that are common to all academic pursuits.

A GOLDEN AGE IN ASTROPHYSICS

The fallout from the failure of the SSC and the resulting revised partnership with government made life at first difficult for particle physicists, and for all physicists. As research funds dried up, competition for the remaining resources increased. The production of new PhDs having just peaked in 1994 following the buildup during the 1980s, many young physicists once again found themselves without jobs. But the skills acquired in earning a PhD in physics, even in the esoteric field of particle physics, could be useful elsewhere in the new economy. Some young physicists turned to computer science or to the high-tech industries where R&D support now far exceeded federal funding. Others turned their analytical skills to work on Wall Street where the stock market was building toward the "dot-com" bubble of the late 1990s. Still others turned to science education, where their skills were, since the 1970s, desperately needed in helping to return American science education to competitive international standards.

Without a next-generation accelerator, many high-energy physicists turned to astrophysics, where the highest energy events in nature are found to occur. Following the launch of the NASA-funded Hubble Space Telescope in 1990 (and its repair in 1993), interest in astrophysics reached new heights just as these new arrivals brought to the field new insights into the underlying particle physics of cosmic events and the origin and evolution of the universe itself.

The momentum had been building in astrophysics for several decades following a series of startling astronomical discoveries beginning in the early 1960s. Drawing upon data gathered by radio telescopes and other ground-based detectors, these discoveries included the first stellar X-ray source, Scorpius X-1 (1962); "quasi-stellar objects," or quasars (1963); the so-called 3K microwave background radiation originating from the Big Bang (1965); and the existence of pulsating objects, or pulsars, emitting intense pulses of radiation at a high rate of frequency (1967).

In addition, the study of fusion reactions occurring in hydrogen bombs and stars during the 1950s resulted in renewed theoretical interest in the possibility of singularities, or mathematical infinities, occurring in the gravitational collapse of massive stars, as first predicted by the Oppenheimer group in 1939. Further research led during the early 1960s to the theoretical confirmation that such an object–what John Wheeler called in 1967 (for the first time in print) a "black hole"—should indeed occur.[41] The quantum properties and behavior of such objects, including the appearance of so-called Hawking radiation, have became topics of research ever since. The stellar object Cygnus X-1, a strong X-ray source discovered in 1964 by an X-ray detecting rocket, was recognized by 1972 as a dual system with one of its members an unseen object of over three solar masses— the first likely candidate for the existence a black hole. But the uncertainty of the measurements could not yet support this conclusion.

These developments of the 1960s and early 1970s led not only to the widespread acceptance of the still debated Big Bang theory of the origin of the universe at that time, but also, with NASA in full swing, to the launch of satellites to detect radiation from stellar objects eventually across the entire electromagnetic spectrum. The data gathered by these dedicated satellites helped launch and maintain what has been called a "golden age" of astrophysics. Beginning with the launch of the Uhuru X-ray detector in 1970, other satellites included the Infrared Astronomy Satellite in 1983, the Cosmic Background Explorer in 1989, the Hubble Space Telescope in 1990, and the Wilkinson Microwave Anisotropy Probe in 2001.

Satellite and ground-based detectors led in turn to greater astrophysical understanding of stars, galaxies, and exotic physical objects such as quasars, pulsars, and black holes. As the Hubble Telescope yielded more and more evidence for the predicted appearance of relativistic jets of elementary particles spewed light-years into space from the accretion disks of apparent black holes, the evidence for the existence of black holes reached a tipping point by the end of the century. The visual evidence was enhanced by the apparent discovery of a supermassive black hole at the center of our Milky Way galaxy. The orbiting Chandra X-Ray Observatory confirmed the discovery in 2004, and the existence of black holes, despite their seemingly impossible infinite densities, was no longer in doubt.

Satellites, together with theoretical advances, have also revealed many more startling features, and puzzling aspects, regarding cosmology, the origin and structure of the universe. In 1979 particle theorist Alan Guth, drawing upon the Standard Model of particle physics, showed that many of the features of the universe, especially its apparently flat spatial structure, could be explained as the result of an enormously rapid acceleration, or "inflation," of the universe during the first tiny fraction of a second following the Big Bang. Satellite studies of the cosmic microwave background confirmed by 2000 that the universe is indeed spatially flat. They also revealed, during the 1990s, the existence of large- and small-scale structure to the background radiation that could account for the origins of the first galaxies and clusters of galaxies in the early universe.[42]

However, studies of the orbital velocities of galaxy clusters by Caltech physicist Fritz Zwicky in 1933, then again by Vera Rubin in the late 1960s, revealed that the visible mass of galaxy clusters was actually not sufficient to prevent the clusters from spinning apart. A large amount of invisible mass, now called "dark matter," had to be present. The mysterious dark matter is now estimated to constitute about 23 percent of the total mass of the universe. The mystery deepened with the discovery in 1998 from observations of supernovae in distant galaxies that the universe is not only expanding outward but that the expansion has been accelerating during a large part of its history. The acceleration appears to arise from the pressure exerted by an unknown "dark energy" permeating the universe and representing as much as 72 percent of the mass of the universe. With ordinary matter making up only about 4.6 percent of the universe's mass, the nature of the other 95 percent of the universe continues to pose a difficult puzzle for particle astrophysicists. It is a puzzle that may be solved through the workings of the world's last remaining superaccelerator, the Large Hadron Collider, or LHC, completed at CERN in 2010.[43]

GLOBALIZING PHYSICS

High-energy physicists who did not leave their field following the demise of the SSC became engaged in increasing international collaboration at home and participation in international research abroad—the LHC in particular. Built in a 17-mile circular tunnel under the Swiss-French border at a cost of $16 billion, the two-ringed collider began in March 2010 to collide beams of protons at record energies of 3.5 TeV (trillion electron volts) in search of the elusive Higgs and other particles that could confirm or refute the reigning Standard Model. Eventually reaching 7 TeV or more (though much less than the projected 20 TeV of the ill-fated SSC), the Large Hadron Collider may point the way to theories beyond the Standard Model, perhaps to the unification of all four of nature's forces at the moment of the Big Bang or to the nature of the mysterious "dark matter" and "dark energy" making up most of the mass in the universe. As of late 2009, the United States had contributed over half a billion dollars to the Large Hadron Collider, and Americans constituted about 1,700 of the 10,000 physicists at work on the giant particle detectors where the collisions occur, the largest national group at the European accelerator.[44]

The trend toward international travel and collaboration has worked both ways since the "brain drain" of the 1960s brought numerous unemployed

European physicists and other scientists to the United States. The trend accelerated greatly after the end of the Cold War. By the end of the century, even more foreign-born physicists were publishing in the leading American physics journals *Physical Review* and *Physical Review Letters*, and more foreign students and researchers were coming to the United States than ever before.

As the United States did earlier in its drive to excellence, other nations, both developed and developing, were pouring more resources into future economic and military development through the funding of science and technology, including physics in particular. A recent study of the global output of scientific publications over time found that in 1995 the twenty-seven nations of the European Union surpassed the United States in the total output of research publications and remained ahead thereafter. The Asian-Pacific nations did the same in 2008. The same report found that in 2010 the output of U.S. research was still concentrated in familiar fashion in a relatively small number of elite research universities that were themselves geographically concentrated on the two coasts and in the upper Midwest.[45]

Most American graduate-level universities welcomed foreign students during the last decades of the century because of the high quality of their preparation and because there were usually not enough American students to fill all of the available research and training programs. At many universities, foreign students and researchers were in the majority. To some, the many languages and cultures represented in laboratories seemed like a miniature United Nations. At Purdue University, known primarily for its science and engineering, the percentage of graduate students from outside the United States jumped from 29 percent in 1991 to 59 percent in 2000–2001.[46]

The Purdue experience agrees with data compiled by the American Institute of Physics. According to the AIP, the total number of physics graduate students in American universities who were foreign citizens rose almost steadily since at least 1975. In 2001 the number of foreign physics students matched the number of U.S. citizens for the first time. A year later, the number of PhDs awarded to foreign students exceeded for the first time those awarded to American citizens. By 2005 the number of PhDs awarded to foreign citizens climbed to a high of 60 percent of all physics doctorates conferred. Within two years, however, this figure dropped back to 55 percent and perhaps lower thereafter, owing to visa restrictions imposed by the Bush administration in the wake of the terror attacks of September 11, 2001.[47]

Paralleling the internationalization of the U.S. physics community, the globalization of physics research occurred in concert with the globalization of the world economy through free-trade agreements, multinational corporations, and the worldwide Internet and Web. The preeminence of the United States and a few other Western nations in a variety of areas including physics no longer goes unchallenged. Still, in 2010 the number of graduate and undergraduate physics students and the number of physics PhDs awarded by American universities, whether to foreign or domestic students, were on the rise, even as other nations kept pace. Membership in the American Physical Society was at an all time high, even as nearly a third of the nonstudent members resided abroad; and, as previously noted, total federal and corporate funding for R&D were also on the rise, although slowed by the economic recession starting in 2008.[48]

Even after the United States relinquished its grandiose claims on the century along with its privileged position in science while it expanded the international identity of its physics discipline during the last third of the century, the nation remained, at the threshold to the twenty-first century, a leading player on the global landscape of physics. But it is doubtful that, a century from now, a similar history focused on physics in a single nation will be possible.

Appendix: Tables

Table 1. Physicists and scientists listed in *American Men of Science*, 1921 and 1938

Year and field	Total	Men			Women			
		Number of men	Physicists as % of male scientists	% with PhD	Number of women	Physicists as % of female scientists	% with PhD	Women as % of field
1921								
Physics	888	864	9.6	79.8	24[a]	5.3	62.5	2.7
All sciences	9,489	9,036		58.2	453		71.8	4.8
1938								
Physics	1,888	1,825	7.2	74.8	63	3.3	73.0	3.3
All sciences	27,287	25,375		70.3	1,912		83.2	7.1

Source: Derived from Margaret W. Rossiter, *Women Scientists in America: Struggles and Strategies to 1940* (Baltimore: Johns Hopkins University Press, 1982), pp. 27, 134, 136, 157.

a. Rossiter found three more women than the twenty-one listed in *AMS* for 1921.

Table 2. Federal R&D outlays for selected years and agencies, in millions of dollars

Fiscal year[a]	Agency	Total R&D[b]	In constant 2000 dollars	Basic research	Applied research[c]	% of all federal outlays	% of federal R&D
1951	All agencies	1,852	12,265	NA	NA	4.1	
	DoD	1,276		NA	NA		68.9
	AEC[d]	278		NA	NA		15.0
	NSF	0.151		NA	NA		<0.1
1955	All agencies	2,252	14,670	130	NA	3.3	
	DoD	1,668		20	NA		74.1
	AEC	294		42	NA		13.1
	NSF	10.3		9.7	NA		0.4
1957	All agencies	4,389	26,896	262	662	5.7	
	DoD	3,250		89	362		74.1
	AEC	645		55	44		14.7
	NSF	38		30	0		0.9
1962	All agencies	11,069	63,115	986	2,018	10.4	
	DoD	6,816		204	1,107		61.6
	AEC	1,316		192	55		11.9
	NASA	1,684		196	250		15.2
	NSF	170		104	0		1.5
1967	All agencies	17,149	88,415	1,846	2,786	10.9	
	DoD	8,136		284	1,307		47.4
	AEC	1,485		302	90		8.7
	NASA	4,988		328	471		29.1
	NSF	328		239	2		1.9
1970	All agencies	15,863	70,402	1,926	2,975	8.1	
	DoD	7,501		317	1,012		47.3
	AEC	1,612		287	146		10.2
	NASA	3,833		358	673		24.2
	NSF	312		245	30		2.0
1974	All agencies	18,176	63,487	2,388	3,788	6.8	
	DoD	8,590		303	1,131		47.3
	Energy R&D	1,882		270	195		10.4
	NASA	3,101		306	640		17.1
	NSF	568		415	105		3.1
1980	All agencies	31,386	65,386	4,674	6,923	5.3	
	DoD	14,189[b]		540	1,721		45.2
	DoE	5,778		523	754		18.4
	NASA	3,393		559	1,051		10.8
	NSF	900		815	58		2.9

Table 2 (continued)

Fiscal year[a]	Agency	Total R&D[b]	In constant 2000 dollars	Basic research	Applied research[c]	% of all federal outlays	% of federal R&D
1985	All agencies	50,180	80,307	7,819	8,315	5.3	
	DoD	30,322		861	2,307		60.4
	DoE	5,834		943	1,198		11.6
	NASA	3,662		751	1,033		7.3
	NSF	1,419		1,262	84		2.8
1989	All agencies	63,572	88,283	10,602	10,164	5.6	
	DoD	38,076		948	2,708		59.9
	DoE	6,066		1,411	1,021		9.5
	NASA	5,913		1,417	1,461		9.3
	NSF	1,724		1,563	108		2.7
1995	All agencies	70,443	79,443	13,877	14,557	4.7	
	DoD	33,857		1,248	2,950		48.1
	DoE	6,890		1,634	1,826		9.8
	NASA	9,640		1,978	2,068		13.7
	NSF	2,439		1,973	176		3.5
2000	All agencies	77,356	77,356	19,570	18,901	4.3	
	DoD	33,215		1,230	3,690		42.9
	DoE	6,874		2,176	1,925		8.9
	NASA	9,755		2,305	1,659		12.6
	NSF	2,942		2,540	186		3.8

Sources: National Science Foundation, "Table A: Federal Obligations for Research and Development, by Character of Work, R&D Plant, and Major Agency: Fiscal Years 1951–2002," at www.nsf.gov/statistics; and U.S. Office of Budget and Management, *Historical Tables, Budget of the U.S. Government, Fiscal Year 2011,* Table 1.1, at www.gpoaccess.gov/usbudget/fy11/pdf/hist.pdf.

Abbreviations: AEC, Atomic Energy Commission; DoD, Department of Defense; DoE, Department of Energy; NASA, National Aeronautics and Space Administration; NSF, National Science Foundation; R&D, research and development.

a. Until fiscal year (FY) 1977, the FY ran from July 1 to June 30 of the next calendar year; after that it ran from October 1 to September 30. The calendar year of the latter date defined the FY number.

b. Includes R&D Plant.

c. Funds for Development and R&D Plant are not shown.

d. In 1974, the AEC's R&D mandate transferred to the Energy Research and Development Agency. In 1977, the Department of Energy assumed this task.

Table 3. American physicists and astronomers and all scientists and engineers by
gender, 1955–1970

Year	Field	Total	Men	Women	Women as % of field
1955	Physics and astronomy	11,452	11,104	348	3.0
	Scientists and engineers	115,775	108,063	7,712	6.7
1960	Scientists and engineers	201,292	187,741	13,551	6.7
1966	Scientists and engineers	242,763	222,599	20,164	8.3
1970	Physics and astronomy	36,336	34,982	1,354	3.7
	Scientists and engineers	312,644	283,351	29,293	9.4

Source: Derived from Margaret W. Rossiter, *Women Scientists in America: Before Affirmative Action,
1940–1972* (Baltimore: Johns Hopkins University Press, 1995), 98, 100. Based on National Science
Foundation, *American Science Manpower,* annual.
   Note: Data on physics and astronomy for 1960 and 1966 are not available.

Table 4. Doctorates awarded in physics and all sciences and engineering by gender of
recipient for selected years, 1966–2000

Year	Field	Total	Men	Women	Women as % of field
1966	Physics	995	976	19	1.9
	All science and engineering	11,570	10,646	924	8.0
1971	Physics	1,625[a]	1,577	48	3.0
	All science and engineering	19,381	17,385	1,996	10.3
1980	Physics	862	808	54	6.3
	All science and engineering	17,775	13,814	3,961	22.3
1990	Physics	1,265	1,135	130	10.3
	All science and engineering	22,868	16,498	6,370	27.9
2000	Physics	1,205	1,039	163	13.5
	All science and engineering	25,921	16,525	9,396	36.2

Source: National Science Foundation, *Science and Engineering Degrees: 1966–2000, Detailed Statistical
Tables* (2000), at www.nsf.gov/statistics/nsf02327, tables 1, 3, 38.
   a. High for the century.

Table 5. Federal and industrial R&D funding for selected years, in millions
of dollars

Year	Federal[a]		Industrial	
	Current dollars	Constant 2000 dollars	Current dollars	Constant 2000 dollars
1953	2,167	13,976	2,200	14,189
1955	2,252	14,670	2,460	15,806
1962	11,069	63,115	5,029	28,675
1967	17,149	88,415	8,020	41,349
1970	15,863	70,402	10,288	45,660
1974	18,176	63,487	14,667	51,230
1980	31,386	65,386	30,476	63,689
1981	34,590	65,527	35,428	67,114
1985	50,180	80,307	57,043	91,290
1989	63,572	88,283	73,501	102,072
1995[b]	70,443	79,443	108,652	122,768
2000	77,356	77,356	180,421	180,421

Sources: National Science Foundation, "Table A: Federal Obligations for Research and
Development, by Character of Work, R&D Plant, and Major Agency: Fiscal Years 1951–2002,"
at www.nsf.gov/statistics; NSF, "Table A-1: Trends in Total Funds for Industrial R&D
Performance in the U.S. by Source of Funds, 1953–2000," at www.nsf.gov/statistics.

a. Includes R&D Plant.

b. Because of a revised sample design, industrial statistics after 1991 are not directly
comparable with those for earlier years.

Notes

INTRODUCTION

1. Henry R. Luce, "The American Century," *Life*, February 17, 1941, 61–65, reprinted in *The Ideas of Henry Luce*, ed. John K. Jessup (New York: Atheneum, 1969), 105–120. See Alan Brinkley, *The Publisher: Henry Luce and His American Century* (New York: Knopf, 2010); G. Pascal Zachary, *Endless Frontier: Vannevar Bush, Engineer of the American Century* (Cambridge, MA: MIT Press, 1999). The attitude is depicted by John Krige, *American Hegemony and the Postwar Reconstruction of Science in Europe* (Cambridge, MA: MIT Press, 2006); D. C. Cassidy, *J. Robert Oppenheimer and the American Century* (New York: Pi Press, 2005; repr., Baltimore: Johns Hopkins University Press, 2009); and Michael Hiltzik, *Colossus: Hoover Dam and the Making of the American Century* (New York: Free Press, 2010).

2. See Ellis W. Hawley and Robert E. Kohler, "Essay Reviews: Two Appraisals, Government Science," review of *Science in the Federal Government* (reissue), by A. Hunter Dupree, *Isis* 78, no. 4 (1987): 576–589.

1. ENTERING THE NEW CENTURY

1. *Physical Review* 10, nos. 1–5 (January–May 1900); 11, nos. 1–5 (July–November 1900). See also Morton Hamermesh, "The Early Years," in *The* Physical Review: *The First Hundred Years, A Selection of Seminal Papers and Commentaries*, ed. Henry H. Stroke (Woodbury, NY: American Institute of Physics Press, 1995), 17–22.

2. Robert H. Kargon and Scott G. Knowles, "Knowledge for Use: Science, Higher Learning, and America's Industrial Heartland, 1880–1915," *Annals of Science* 59 (2002): 1–20.

3. Purdue University, *Course Catalogue 1895–1896* (Lafayette, IN: Purdue University, 1895). I thank Prof. Arnold Tubis for sending me copies of this and

other course catalogues. For more on this period at Purdue, see Solomon Garten-haus, Arnold Tubis, and David C. Cassidy, "A History of Physics at Purdue: The Early Years, 1874–1928," Purdue University, College of Science, Department of Physics, www.physics.purdue.edu/about_us/history/early_years.shtml (*Purdue Physics: Past, Present, and Future*, vol. 3 [1994], 9–14).

4. Daniel J. Kevles, *The Physicists: The History of a Scientific Community in Modern America* (New York: Vintage Books, 1971; repr. with a new essay on the Superconducting Super Collider, Cambridge, MA: Harvard University Press, 1995), 26, 39.

5. Ibid., 38.

6. A. Hunter Dupree, *Science in the Federal Government* (Cambridge, MA: Harvard University Press, 1957; repr., Baltimore: Johns Hopkins University Press, 1986), chap. 14.

7. Ibid.

8. Marcia Graham Synnott, "Anti-Semitism and American Universities: Did Quotas Follow the Jews?" in *Anti-Semitism in America*, American Jewish History, vol. 6, part 2, ed. Jeffrey S. Gurock (New York: Routledge, 1998), 236.

9. These developments are further discussed by Dupree, *Science in the Federal Government*, chap. 12, and Kevles, *The Physicists*, chap. 5.

10. See Christophe Lécuyer, "MIT, Progressive Reform, and 'Industrial Service,' 1890–1920," *Historical Studies in the Physical Sciences* 26, no. 1 (1995): 35–88.

11. Dupree, *Science in the Federal Government*, 294–296. Much has been written about the relationship of science and technology. For this period, see Ronald Kline, "Construing 'Technology' as 'Applied Science': Public Rhetoric of Scientists and Engineers in the United States, 1880–1945," *Isis* 86 (1995): 194–221.

12. Daniel S. Greenberg, *The Politics of Pure Science*, 2nd ed. (Chicago: University of Chicago Press, 1999), chap. 1.

13. Don K. Price, *The Scientific Estate* (Cambridge, MA: Belknap Press of Harvard University Press, 1965), 3.

14. Henry A. Rowland, "A Plea for Pure Science," *Science* 2 (August 24, 1883): 242–250, on p. 244, address as vice president of Section B (Physics) of American Association for the Advancement of Science (AAAS), August 15, 1883.

15. The German outlook is further discussed by Fritz K. Ringer, *The Decline of the German Mandarins: The German Academic Community, 1890–1933* (Cambridge: Cambridge University Press, 1969); and D. Cassidy, *Beyond Uncertainty* (New York: Bellevue Literary Press, 2009), chap. 1. For more on the German influence on the American outlook see David Cahan, "Helmholtz and the Shaping of the American Physics Elite in the Gilded Age," *Historical Studies in the Physical and Biological Sciences* 35, no. 1 (2004): 1–34.

16. Rowland, "Plea for Pure Science."

17. Paul Hartman, *A Memoir on the* Physical Review: *A History of the First Hundred Years* (Woodbury, NY: American Institute of Physics, 1994).

18. Henry A. Rowland, "The Highest Aim of the Physicist," American Physical Society, *Bulletin* 1 (1899): 4–16, on p. 4, www.aip.org/history/gap/Rowland/Rowland.html. For more on Rowland and pure science, see Kevles, *The Physicists*, 43–48.

19. See Leonard S. Reich, *The Making of American Industrial Research: Science and Business at GE and Bell, 1876–1926* (New York: Cambridge University Press, 1985).

20. Quoted by George E. Folk, *Patents and Industrial Progress* (New York: Harper and Bros., 1942), 153.

21. Reich, *American Industrial Research*, introduction.

22. Spencer R. Weart, "The Physics Business in America, 1919–1940: A Statistical Reconnaissance," in *The Sciences in the American Context: New Perspectives*, ed. Nathan Reingold (Washington, DC: Smithsonian Institution Press, 1979), 295–358, tables on pp. 333–334.

23. Reich, *American Industrial Research*, 2. For an early history of Bell Labs, see Lillian Hoddeson, "The Emergence of Basic Research in the Bell Telephone System, 1875–1915," *Technology and Culture* 22 (1981): 512–544.

24. Oliver E. Buckley, "Frank Baldwin Jewett 1879–1949," National Academy of Sciences, *Biographical Memoirs* 27 (1952): 237–264, on p. 259.

25. Quoted by Reich, *American Industrial Research*, 159.

26. Ibid., 159–160.

27. Quoted by Kevles, *The Physicists*, 100.

28. This is argued by Kevles, *The Physicists*, 101.

29. Gartenhaus, Tubis, and Cassidy, "History of Physics at Purdue."

30. Kevles, *The Physicists*, 79.

31. Paul Forman, John L. Heilbron, and Spencer Weart, "Physics *circa* 1900: Personnel, Funding, and Productivity of the Academic Establishments," *Historical Studies in the Physical Sciences* 5 (1975): 1–185, tables on pp. 12 and 31.

2. AMERICAN PHYSICS COMES OF AGE

1. Felix Adler, quoted by Howard B. Radest, *Toward Common Ground: The Story of the Ethical Societies in the United States* (New York: Ungar, 1969), 191–192.

2. Samuel Eliot Morison, ed., *The Development of Harvard University since the Inauguration of President Eliot, 1869–1929* (Cambridge, MA: Harvard University Press, 1930), 459; George Ellery Hale, quoted by Helen Wright, *Explorer of the Universe: A Biography of George Ellery Hale* (Woodbury, NY: AIP Press, 1966; repr., 1994), 288.

3. Stanley Coben, "The Scientific Establishment and the Transmission of Quantum Mechanics to the United States, 1919–32," *American Historical Review* 76 (1971): 442–466.

4. For more on this see Wright, *Explorer.*

5. Hale and scientist-politicians are further discussed by Daniel J. Kevles, *The Physicists: The History of a Scientific Community in Modern America* (New York: Vintage Books, 1971; repr. Cambridge, MA: Harvard University Press, 1995), chaps. 8–9; A. Hunter Dupree, *Science in the Federal Government* (Cambridge, MA: Harvard University Press, 1957; repr., Baltimore: Johns Hopkins University Press, 1986), chap. 16; and Daniel J. Kevles, "George Ellery Hale, the First World War, and the Advancement of Science in America," *Isis* 59 (1968): 427–437.

6. Quoted by Kevles, *The Physicists*, 112.

7. Excerpt from report of NRC meeting, September 1916, quoted by Dupree, *Science in the Federal Government*, 310.

8. Colonel R. J. Burt to Chief of Staff, June 18, 1918, quoted by Kevles, *The Physicists*, 132.

9. Dupree, *Science in the Federal Government*, 314.

10. Robert A. Millikan, *Autobiography* (New York: Prentice-Hall, 1950), 178, also 165, 168.

11. For further discussion of this research, see Kevles, *The Physicists*, chap. 9.

12. Ibid., 138.

13. Dupree, *Science in the Federal Government*, 337.

14. Spencer R. Weart, "The Physics Business in America, 1919–1940: A Statistical Reconnaissance," in *The Sciences in the American Context: New Perspectives*, ed. Nathan Reingold (Washington, DC: Smithsonian Institution Press, 1979), 295–358, esp. p. 296, fig. 1; and from data provided by M. Lois Marckworth, *Dissertations in Physics: An Indexed Bibliography of All Doctoral Theses Accepted by American Universities, 1861–1959* (Stanford, CA: Stanford University Press, 1961). Rossiter found that during the course of the nineteenth century a total of fifty-six women received doctorates in science, including three in physics; Margaret W. Rossiter, *Women Scientists in America: Struggles and Strategies to 1940* (Baltimore: Johns Hopkins University Press, 1982), 36.

15. Quoted in J. L. Heilbron and Robert W. Seidel, *Lawrence and His Laboratory: A History of the Lawrence Berkeley Laboratory*, vol. 1 (Berkeley: University of California Press, 1989), 12.

16. For more on Millikan's research, see Robert H. Kargon, "The Conservative Mode: Robert A. Millikan and the Twentieth-Century Revolution in Physics," *Isis* 68 (1977): 509–526. For more on Millikan, see his *Autobiography* and Robert H. Kargon, *The Rise of Robert Millikan: Portrait of a Life in American Science* (Ithaca, NY: Cornell University Press, 1982).

17. Robert A. Millikan, "Some Exceptional Opportunities in Southern California," application to General Education Board of the Rockefeller Foundation, 11 April 1924. Millikan Papers, box 27, folder 9. California Institute of Technology Archives, Pasadena, CA.

18. Ibid.

19. Hans Bethe, interview by Judith Goodstein, 17 February 1982. California Institute of Technology Archives, Pasadena, CA.

20. Marcia Graham Synnott, "Anti-Semitism and American Universities: Did Quotas Follow the Jews?" in *Anti-Semitism in America*, American Jewish History, vol. 6, part 2, ed. Jeffrey S. Gurock (New York: Routledge, 1998), 233–271, on p. 236.

21. These developments are further discussed by Coben, "Scientific Establishment"; Gerald Holton, "On the Hesitant Rise of Quantum Physics Research in the United States," in *Thematic Origins of Scientific Thought: Kepler to Einstein*, rev. ed. (Cambridge, MA: Harvard University Press, 1988), 147–185; and John W. Servos, "Mathematics and the Physical Sciences in America, 1880–1930," *Isis* 77 (1986): 611–629.

22. See Christa Jungnickel and Russell McCormmach, *Intellectual Mastery of Nature: Theoretical Physics from Ohm to Einstein*, 2 vols. (Chicago: University of Chicago Press, 1986).

23. Bridgman to Kemble, 16 March 1919, quoted by Holton, "Hesitant Rise," 171–172. For more on the origins and growth of theoretical physics in the United States, see Silvan S. Schweber, "The Empiricist Temper Regnant: Theoretical Physics in the United States 1920–1950," *Historical Studies in the Physical Sciences* 17, no. 1 (1986): 55–98.

24. Attributed to Wickliffe Rose, an official of the Rockefeller Foundation, by Raymond B. Fosdick, *Adventure in Giving: The Story of the General Education Board, A Foundation Established by John D. Rockefeller* (New York: Harper and Row, 1962), 230. For more on the NRC strategy regarding the peak universities, postdoctoral fellows, and the importing of quantum mechanics, see Robert E. Kohler, *Partners in Science: Foundations and Natural Scientists, 1900–1945* (Chicago: University of Chicago Press, 1991); Kevles, *The Physicists*, chaps. 13–14; Coben, "Scientific Establishment"; Dupree, *Science in the Federal Government*, chap. 17; Holton, "Hesitant Rise"; and D. C. Cassidy, *J. Robert Oppenheimer and the American Century* (New York: Pi Press, 2005; repr., Baltimore: Johns Hopkins University Press, 2009), chaps. 7–8.

25. Weart, "Physics Business," 298; see also Cassidy, *J. Robert Oppenheimer*, appendix 1.

26. Weart, "Physics Business," 299.

27. Werner Heisenberg, *The Physical Principles of the Quantum Theory*, trans. Carl Eckart and Frank C. Hoyt (Chicago: University of Chicago Press, 1930); Paul Dirac, *Principles of Quantum Mechanics* (New York: Oxford University Press, 1930).

28. Kevles, *The Physicists*, 211–212.

29. Letter of Max Born to dean of science faculty, University of Göttingen, 13 April 1926. Max Born personnel file. University Archive, University of Göttingen, Göttingen, Germany.

30. See esp. Wolfgang Pauli, *Scientific Correspondence with Bohr, Einstein, Heisenberg and Others*, vol. 1: *1919–1929*, ed. A. Hermann et al. (New York: Springer-Verlag, 1979), in the original languages.

31. E. Schrödinger, "An Undulatory Theory of the Mechanics of Atoms and Molecules," *Physical Review* 28 (1926): 1049–1070, dated September 3, 1926. See Paul Hartman, *A Memoir on the* Physical Review: *A History of the First Hundred Years* (Woodbury, NY: American Institute of Physics, 1994), 139. I thank Martin Blume for bringing this to my attention.

32. Weart, "Physics Business," 299.

33. Yves Gingras, "The Transformation of Physics from 1900 to 1945," *Physics in Perspective* 12 (2010): 248–265, fig. 4.

34. Coben, "The Scientific Establishment," 458.

35. Spencer Weart, "The Last Fifty Years—A Revolution?" *Physics Today* 34, no. 11 (November 1981): 37–49, on p. 39.

36. Margaret W. Rossiter, *Women Scientists in America: Struggles and Strategies to 1940* (1982), and *Women Scientists in America: Before Affirmative Action, 1940–1972* (Baltimore: Johns Hopkins University Press, 1995). For accounts of twentieth-century women physicists, see Nina Byers and Gary Williams, *Out of the Shadows: Contributions of Twentieth Century Women to Physics* (New York: Cambridge University Press, 2006).

37. Catherine Hill, Christianne Corbett, and Andresse St. Rode, *Why So Few? Women in Science, Technology, Engineering, and Mathematics* (Washington, DC: American Association of University Women, 2010); and Rossiter, *Women Scientists in America: Struggles and Strategies to 1940*, esp. tables on pp. 134 and 136.

38. "Scientific Careers for Women," *New York Times*, June 4, 1921, 2, further discussed by Rossiter, *Women Scientists in America: Struggles and Strategies to 1940*, chap. 5.

39. Rossiter, *Women Scientists in America: Struggles and Strategies to 1940*, table on p. 27.

40. Ibid., 154.

41. Robert Singerman, "The Jew as Racial Alien: The Genetic Component of American Anti-Semitism," in *Anti-Semitism in American History*, ed. David A. Gerber (Urbana: University of Illinois Press, 1986), 103–128.

42. Elizabeth Bogen, *Immigration in New York* (New York: Praeger, 1987); and Thomas Kessner, *The Golden Door: Italian and Jewish Immigrant Mobility in New York City, 1880–1915* (New York: Oxford University Press, 1977), 48–49.

43. Synnott, "Anti-Semitism," 236.

44. Ibid., 238.

45. Unless otherwise noted, the data cited in this section derive from Synnott, "Anti-Semitism."

46. See Marcia Graham Synnott, *The Half-Opened Door: Discrimination and Admissions at Harvard, Yale, and Princeton, 1900–1970* (Westport, CT: Greenwood Press, 1979), chaps. 2–5.

47. "Lowell Tells Jews Limit at Colleges Might Help Them," *New York Times*, June 17, 1922, 1.

48. Millikan to Richard Tolman, 31 July 1945. Millikan Papers, box 28, folder 15. California Institute of Technology Archives, Pasadena, CA.

49. Biographies of Oppenheimer abound. The following draws mainly on Cassidy, *J. Robert Oppenheimer*, chaps. 6–8.

50. William T. Ham, "Harvard Student Opinion on the Jewish Question," *The Nation*, vol. 115, September 6, 1922, 225–227.

51. Bridgman to Rutherford, 24 June 1925, quoted in Alice Kimball Smith and Charles Weiner, eds., *Robert Oppenheimer: Letters and Recollections* (Cambridge, MA: Harvard University Press, 1980), 77.

52. A bibliography is offered by Smith and Weiner, *Robert Oppenheimer*.

53. Karl T. Compton to Augustus Trowbridge, 6 December 1926. IEB series 1.3, folder 1095. Rockefeller Archives Center, Tarrytown, NY.

54. Oppenheimer to Elmer Hall, chair of Berkeley physics department, 7 March 1928. Raymond T. Birge Papers, MSS 87/147C. Bancroft Library, Berkeley, CA; and NRC, minutes of meeting on 26 April 1928. IEB series 1.3, folder 1095, Rockefeller Archives Center, Tarrytown, NY.

55. Isidor Rabi, interview by Thomas S. Kuhn, 8 December 1963; and J. Robert Oppenheimer, interview by Charles Weiner, November 1966; both in Niels Bohr Library and Archives, American Institute of Physics, College Park, MD.

56. See Lillian Hoddeson, Gordon Baym, and Michael Eckert, "The Development of the Quantum Mechanical Theory of Metals, 1926–1933," in *Out of the*

Crystal Maze: Chapters from the History of Solid-State Physics, ed. Lillian Hoddeson, Ernest Braun, Jürgen Teichmann, and Spencer Weart (New York: Oxford University Press, 1992), 88–181.

57. J. Robert Oppenheimer, interview by Thomas S. Kuhn, 20 November 1963. Niels Bohr Library and Archives, American Institute of Physics, ,College Park, MD.

58. Birge to Lawrence, 23 February 1928. Raymond T. Birge Papers, MSS 87/147C. Bancroft Library, Berkeley, CA.

59. John H. Van Vleck, "American Physics Comes of Age," *Physics Today* 17 (June 1964): 21–26.

60. Kevles, *The Physicists*, 220.

61. John C. Slater, "Quantum Physics in America between the Wars," *Physics Today* 21 (January 1968): 43–51, on p. 43.

3. SURVIVING THE DEPRESSION

1. Quoted in J. L. Heilbron and Robert W. Seidel, *Lawrence and His Laboratory: A History of the Lawrence Berkeley Laboratory*, vol. 1 (Berkeley: University of California Press, 1989), 88.

2. Luis W. Alvarez, "Ernest O. Lawrence, 1901–1958," in *Biographical Memoirs of the National Academy of Sciences*, vol. 45 (Washington, DC: National Academy of Sciences, 1970), 250–294.

3. Accelerator designs through Lawrence's 1.25 MeV machine are discussed by Heilbron and Seidel, *Lawrence and His Laboratory*, chap. 2.

4. *Physical Review*, volumes 35–36 (1930).

5. There is a large literature on the history of nuclei, particles, and quantum fields. Selected works include Abraham Pais, A. B. Pippard, and Laurie M. Brown, eds., *Twentieth Century Physics*, 3 vol. (Bristol and Philadelphia: IOP Press; New York: AIP Press, 1995); Silvan S. Schweber, *QED and the Men Who Made It: Dyson, Feynman, Schwinger, and Tomonaga* (Princeton, NJ: Princeton University Press, 1994); Abraham Pais, *Inward Bound: Of Matter and Forces in the Physical World* (New York: Oxford University Press, 1986); Laurie Brown and Lillian Hoddeson, eds., *The Birth of Particle Physics* (New York: Cambridge University Press, 1983); Peter Galison, *Image and Logic: A Material Culture of Microphysics* (Chicago: University of Chicago Press, 1997); Peter Galison, *How Experiments End* (Chicago: University of Chicago Press, 1987); Stephen Weinberg, *The Discovery of Subatomic Particles*, 2nd ed. (New York: Cambridge University Press, 2003).

6. Hans A. Bethe, "J. Robert Oppenheimer, 1904–1967," *Biographical Memoirs of Fellows of the Royal Society* [London] 14 (1968): 391–416, on p. 396.

7. Joel Lebowitz, Wolfgang Panofsky, and Stuart Rice, "Melba Newell Phillips," *Physics Today* 58 (July 2005): 80–81.

8. Daniel J. Kevles, *The Physicists: The History of a Scientific Community in Modern America* (New York: Vintage Books, 1971; repr. Cambridge, MA: Harvard University Press, 1995), 250; A. Hunter Dupree, *Science in the Federal Government* (Cambridge, MA: Harvard University Press, 1957; repr., Baltimore: Johns Hopkins University Press, 1986), 346.

9. D. C. Cassidy, *J. Robert Oppenheimer and the American Century* (New York: Pi Press, 2005; repr., Baltimore: Johns Hopkins University Press, 2009), 138, derived from department records.

10. Robin Rider, "Alarm and Opportunity: Emigration of Mathematicians and Physicists to Britain and the United States, 1933–1945," *Historical Studies in the Physical Sciences* 15 (1984): 107–176, esp. pp. 154 and 160.

11. See Edward Regis, *Who Got Einstein's Office: Eccentricity and Genius at the Institute for Advanced Study* (Reading, MA: Addison-Wesley, 1987).

12. Marcia Graham Synnott, *The Half-Opened Door: Discrimination and Admissions at Harvard, Yale, and Princeton, 1900–1970* (Westport, CT: Greenwood Press, 1979), and "Anti-Semitism and American Universities: Did Quotas Follow the Jews?" in *Anti-Semitism in America*, American Jewish History, vol. 6, part 2, ed. Jeffrey S. Gurock (New York: Routledge, 1998), 233–271.

13. Goudsmit to Randall, 13 July 1938. Goudsmit Papers, letter files, Niels Bohr Library and Archives, American Institute of Physics, , College Park, MD; and Einstein to Pauli, September 1938, in Wolfgang Pauli, *Wissenschaftlicher Briefwechsel*, vol. 2, ed. Karl von Meyenn et al. (Berlin: Springer-Verlag, 1985), 600–601.

14. *Newsweek*, November 7, 1936, 29, quoted by Kevles, *The Physicists*, 282.

15. Discussed by Dupree, *Science in the Federal Government*, 347–349; Kevles, *The Physicists*, chap. 16.

16. Robert E. Kohler, *Partners in Science: Foundations and Natural Scientists, 1900–1945* (Chicago: University of Chicago Press, 1991), chap. 10; Kevles, *The Physicists*, 248.

17. Heilbron and Seidel, *Lawrence and His Laboratory*, 180, tables 5.1 and 6.4; Robert Seidel, "The Origins of the Lawrence Berkeley Laboratory," in *Big Science: The Growth of Large-Scale Research*, ed. Peter Galison and Bruce W. Hevly (Stanford, CA: Stanford University Press, 1992), 21–45; Alvarez, "Ernest O. Lawrence," 258.

18. Karl T. Compton, "The Government's Responsibilities in Science," *Science* 81 (1935), 347–355, on p. 349.

19. Further described by Kevles, *The Physicists*, 357–359.

20. Robert A. Millikan, excerpt from radio address, in "Millikan Scores Federal Meddling," *New York Times*, August 7, 1934, 1.

21. Julius A. Stratton, "Karl Taylor Compton, 1887–1954," in *Biographical Memoirs of the National Academy of Sciences*, vol. 61 (Washington, DC: National Academy of Sciences, 1992), 37–57, on p. 45.

22. Listed by Compton, "Government's Responsibilities," 347–348.

23. Quoted by Kevles, *The Physicists*, 257.

24. Further discussed by Dupree, *Science in the Federal Government*, 350–358.

25. Science Committee of National Resources Board, *Research—A National Resource: I—Relation of Federal Government to Research* (Washington, DC: Government Printing Office, 1938).

26. Quoted by Kevles, *The Physicists*, 264.

27. Compton, "Government's Responsibilities," 354.

28. Patrick J. McGrath, *Scientists, Business and the State, 1890–1960* (Chapel Hill: University of North Carolina Press, 2001), chap. 2.

29. Dupree, *Science in the Federal Government*, 366.

30. Solomon Gartenhaus, Arnold Tubis, and D. C. Cassidy, "A History of Physics at Purdue: Early Lark-Horovitz Era (1928–1942)," Purdue University, College of Science, Department of Physics, www.physics.purdue.edu/about_us/history/early_years.shtml (repr., *Purdue Physics: Past, Present, and Future*, vol. 4, 1995, no. 1, 5–12); Silvan S. Schweber, "Big Science in Context: Cornell and MIT," in Galison and Hevly, *Big Science*, 149–183, esp. pp. 159–165.

31. Raymond T. Birge, report to Berkeley physics department faculty meeting, 24 August 1938. Records of the Department of Physics, CU-68, box 2, folder 2:4. Bancroft Library, Berkeley, CA.

32. Silvan S. Schweber, "The Empiricist Temper Regnant: Theoretical Physics in the United States 1920–1950," *Historical Studies in the Physical Sciences* 17, no. 1 (1986): 55–98, on 84.

33. Quoted by Kevles, *The Physicists*, 285.

34. Kohler, *Partners in Science*, 371–375; Seidel, "The Origins," 28, found that 40 percent of the funding for Lawrence's accelerators through 1940 came from the state, 38 percent from foundations, and 22 percent from the federal government.

35. Heilbron and Seidel, *Lawrence and His Laboratory*, 310, table 6.5.

36. Ibid., 212, table 5.1.

37. *Time* magazine, 1 November 1937, cover.

38. See the works cited in note 5, and Peter Galison, "The Discovery of the Muon and the Failed Revolution against Quantum Electrodynamics," *Centaurus* 26 (1983): 262–316; and D. C. Cassidy, "Cosmic Ray Showers, High Energy Physics, and Quantum Field Theories: Programmatic Interactions in the 1930s," *Historical Studies in the Physical Sciences* 12, no. 1 (1981): 187–224.

39. Planning for the machine and maneuvering for its funds are discussed by Heilbron and Seidel, *Lawrence and His Laboratory*, 471–484.

4. THE PHYSICISTS' WAR

1. Albert Einstein, letter to F. D. Roosevelt, August 2, 1939, in Albert Einstein, *Einstein on Peace*, ed. Otto Nathan and Heinz Norden (New York: Schocken Books, 1960), 294–295. According to the editors, the letter was written by Leo Szilard but based on an earlier German draft by Einstein, ibid., 292–294.

2. The work of the science administrators leading up to the Manhattan Project is further discussed by, among others, Richard Rhodes, *The Making of the Atomic Bomb* (New York: Simon and Schuster, 1986), chap. 13; Daniel J. Kevles, *The Physicists: The History of a Scientific Community in Modern America* (New York: Vintage Books, 1971; repr., Cambridge, MA: Harvard University Press, 1995), chap. 19; Nathan Reingold, "Vannevar Bush's New Deal for Research; or, The Triumph of the Old Order," in *Science, American Style* (New Brunswick, NJ: Rutgers University Press, 1991), 284–333; D. C. Cassidy, *J. Robert Oppenheimer and the American Century* (New York: Pi Press, 2005; repr., Baltimore: Johns Hopkins University Press, 2009), chaps. 11–12.

3. Jerome B. Wiesner, "Vannevar Bush, 1890–1974," *Biographical Memoirs of the National Academy of Sciences*, vol. 50 (Washington, DC: National Academy of Sciences, 1979), 87–117, on p. 89. See also G. Pascal Zachary, *Endless Frontier: Vannevar Bush, Engineer of the American Century* (Cambridge, MA: MIT Press, 1999).

4. Wiesner, "Vannevar Bush."

5. The social and political views of scientists during the late 1930s and early 1940s are discussed by Peter J. Kuznick, *Beyond the Laboratory: Scientists as Political Activists in 1930s America* (Chicago: University of Chicago Press, 1987).

6. See the works in note 2.

7. Carroll Pursell, "Science Agencies in World War II: The OSRD and Its Challengers," in *The Sciences in the American Context: New Perspectives*, ed. Nathan Reingold (Washington, DC: Smithsonian Institution Press, 1979), 359–378.

8. Kevles, *The Physicists*, 342. The top five in dollar amounts were MIT, Caltech, Harvard, Columbia, and University of California, Berkeley (Pursell, "Science Agencies," 364).

9. Kevles, *The Physicists*, 320; and Department of Physics, UC Berkeley, report for 1940–1942, dated 1 May 1942. Records of the Department of Physics (archive collection), CU-68, box 2, folder 7. Bancroft Library, Berkeley, CA. Kevles estimated that the 1,700 represented about one quarter of the physics profession and three quarters of its leadership.

10. "The MAUD Report, 1941," in *The American Atom: A Documentary History of Nuclear Policies from the Discovery of Fission to the Present, 1939–1984*, ed. Robert C. Williams and Philip L. Cantelon (Philadelphia: University of Pennsylvania Press, 1984), 19–23, quoting from pp. 20 and 23.

11. FDR to Bush, 11 March 1942, quoted by Patrick J. McGrath, *Scientists, Business, and the State, 1890–1960* (Chapel Hill: University of North Carolina Press, 2002), 213n7, FDR's emphasis.

12. Bush to Compton, 13 December 1941, Bush-Conant File Relating to the Development of the Atomic Bomb [microform], record group 227, roll 1, frame 0498 (National Archives II, College Park, MD). On Lawrence and the separation of isotopes, see J. L. Heilbron and Robert W. Seidel, *Lawrence and His Laboratory: A History of the Lawrence Berkeley Laboratory*, vol. 1 (Berkeley: University of California Press, 1989), 515–517.

13. Bush to Conant, 9 October 1941, Bush-Conant File, note 12, roll 1, frame 0529.

14. Bush to Compton, 9 October 1941, ibid., frame 0401.

15. Alex Roland, "Science and War," in *Historical Writing on American Science: Perspectives and Prospects*, ed. Sally Gregory Kohlstedt and Margaret Rossiter (Baltimore: Johns Hopkins University Press, 1986), 247–272, on p. 264.

16. Kevles, *The Physicists*, 320.

17. Histories/discussions of the Rad Lab and radar work are offered by Henry E. Guerlac, *Radar in World War II*, 2 vols. (New York: AIP Press, 1987); Kevles, *The Physicists*, chap. 20; Silvan S. Schweber, *QED and the Men Who Made It: Dyson, Feynman, Schwinger, and Tomonaga* (Princeton, NJ: Princeton University Press, 1994), 136–141; and Heilbron and Seidel, *Lawrence and His Laboratory*, 494–501.

18. Schweber, *QED*, 137; Kevles, *The Physicists*, 304.

19. Heilbron and Seidel, *Lawrence and His Laboratory*, 496, table 10.1.

20. About $27 billion in 2010 dollars (the inflation adjustment is based on the consumer price index calculator provided by the U.S. Bureau of Labor Statistics at www.bls.gov/data/inflation_calculator.htm). Also see Kevles, *The Physicists*, 307; Schweber, *QED*, 137.

21. Kevles, *The Physicists*, 308.

22. Further described by Heilbron and Seidel, *Lawrence and His Laboratory*, 510–517.

23. Lawrence to Conant, 26 March 1942, Bush-Conant File, note 12, roll 1, frame 0714.

24. The best general history of the building of the atomic bomb is offered by Rhodes, *Making of the Atomic Bomb*. Accounts of the technical work include Lillian Hoddeson, Paul W. Henriksen, Roger A. Meade, and Catherine L. Westfall, *Critical Assembly: A Technical History of Los Alamos during the Oppenheimer Years, 1943–1945* (New York: Cambridge University Press, 1993); and Henry deWolf Symth, *Atomic Energy for Military Purposes: The Official Report on the Development of the Atomic Bomb under the Auspices of the United States Government, 1940–1945* (Princeton, NJ: Princeton University Press, 1946). See also Peter Galison, *Image and Logic: A Material Culture of Microphysics* (Chicago: University of Chicago Press, 1997), chap. 4.

25. Quoted by Alice Kimball Smith and Charles Weiner, eds., *Robert Oppenheimer: Letters and Recollections* (Cambridge, MA: Harvard University Press, 1980), 227.

26. Robert S. Norris, *Racing for the Bomb: General Leslie R. Groves, the Manhattan Project's Indispensable Man* (South Royalton, VT: Steerforth Press, 2002), esp. 240–243 on the selection of Oppenheimer.

27. Bush to Conant, 18 January 1943, Bush-Conant File, note 12, roll 3, frame 0767; Rhodes, *Making of the Atomic Bomb*, 451.

28. Their stories are offered in Cynthia C. Kelly, ed., *The Manhattan Project: The Birth of the Atomic Bomb in the Words of Its Creators, Eyewitnesses, and Historians* (New York: Black Dog & Leventhal, 2007). See also Ruth H. Howes and Caroline C. Herzenberg, *Their Day in the Sun: Women of the Manhattan Project* (Philadelphia, PA: Temple University Press, 2003).

29. See Lillian Hoddeson et al., *Critical Assembly*.

30. Rhodes, *Making of the Atomic Bomb*, 478–480.

31. Oppenheimer to Groves, 6 October 1944, and Conant to Oppenheimer, 20 October 1944, both in Smith and Weiner, *Robert Oppenheimer*, 286–287.

32. Charles Thorpe, *Oppenheimer: The Tragic Intellect* (Chicago: University of Chicago Press, 2008). Physicists' growing awareness of the issue of social responsibility is explored by Lawrence Badash, "American Physicists, Nuclear Weapons in World War II, and Social Responsibility," *Physics in Perspective* 7 (2005), 138–149.

33. Joseph Rotblat, "Leaving the Bomb Project," *Bulletin of the Atomic Scientists* 41 (August 1985): 15–19, on p. 18, excerpted by Kelly, *Manhattan Project*, 279–282, on p. 280.

34. "Chicago Scientists' Petition to the President, July 17, 1945," published in Williams and Cantelon, *American Atom*, 66–67.

35. "The Franck Report, June 11, 1945," in Alice Kimball Smith, *A Peril and a Hope: The Scientists' Movement in America, 1945–1947* (Cambridge, MA: MIT Press, 1970), 371–383.

36. Ibid., 371–372.

37. "Science Panel: Recommendations on the Immediate Use of Nuclear Weapons, June 16, 1945," excerpted in Williams and Cantelon, *American Atom*, 63–64.

38. For more on the decision and later reflections, see Barton Bernstein, "Four Physicists and the Bomb: The Early Years 1945–1950," *Historical Studies in the Physical Sciences* 18, no. 2 (1988): 231–263.

39. "Science Panel," Williams and Cantelon, *American Atom*, 64.

40. Quoted by Rhodes, *Making of the Atomic Bomb*, 692.

5. TAMING THE ENDLESS FRONTIER

1. Vannevar Bush, *Science—The Endless Frontier: A Report to the President on a Program for Postwar Scientific Research* (Washington, DC: Government Printing Office, 1945), dated July 1945, p. 18, available at www.nsf.gov/od/lpa/nsf50/vbush1945.htm.

2. Ibid., 34.

3. For more on Bush and the NSF, see Daniel J. Kevles, "The National Science Foundation and the Debate over Postwar Research Policy, 1942–1945: A Political Interpretation of *Science—The Endless Frontier*," *Isis* 68, no. 1 (1977): 5–26. On German physicists and funding, see Brigitte Schröder-Gutehus, "The Argument for the Self-Government and Public Support of Science in Weimar Germany," *Minerva* 10 (1972): 537–570; and Paul Forman, "The Financial Support and Political Alignment of Physicists in Weimar Germany," *Minerva* 12 (1974): 39–66.

4. Daniel J. Kevles, *The Physicists: The History of a Scientific Community in Modern America* (New York: Vintage Books, 1971; repr. Cambridge, MA: Harvard University Press, 1995), 375; National Science Foundation (NSF), "Federal Funds for R&D," NSF 01–334 statistical tables, "Table A. Federal Obligations for Research and Development, by Character of Work, R&D Plant, and Major Agency: Fiscal Years 1951–2002," at www.nsf.gov/statistics/fedfunds.

5. Alfred K. Mann, *For Better or for Worse: The Marriage of Science and Government in the United States* (New York: Columbia University Press, 2000); Robert Seidel, "Accelerating Science: The Postwar Transformation of the Lawrence Berkeley Laboratory," *Historical Studies in the Physical Sciences* 13, no. 2 (1983): 375–400.

6. Harvey M. Sapolsky, *Science and the Navy: The History of the Office of Naval Research* (Princeton, NJ: Princeton University Press, 1990).

7. Kevles, *The Physicists*, 363.

8. DuBridge, Statement to Joint Congressional Committee, 8 July 1949. Lee A. DuBridge papers, box 167, folder 4. California Institute of Technology Archives, Pasadena, CA.

9. This has been a subject of considerable study and with varying conclusions. See, for example, Paul Forman, "Behind Quantum Electronics: National Security as Basis for Physical Research in the United States, 1940–1960," *Historical Studies in the Physical Sciences* 18, no. 1 (1987): 149–229; Daniel Kevles, "Cold War and Hot Physics: Science, Security, and the American State, 1945–1956," *Historical Studies in the Physical Sciences* 20, no. 2 (1990): 239–264; Roger L. Geiger, *Research and Relevant Knowledge: American Research Universities since World War II* (New York: Oxford University Press, 1993); Roger L. Geiger, "Science, Universities, and National Defense, 1945–1970," *Osiris* 7 (1992): 26–48; Harvey M. Sapolsky, "Academic Science and the Military: The Years since the Second World War," in *The Sciences in the American Context: New Perspectives*, ed. Nathan Reingold (Washington, DC: Smithsonian Institution Press, 1979), 379–399; Rebecca S. Lowen, *Creating the Cold War University: The Transformation of Stanford* (Berkeley: University of California Press, 1997); and Stuart W. Leslie, *The Cold War and American Science: The Military-Industrial-Academic Complex at MIT and Stanford* (New York: Columbia University Press, 1993).

10. Forman, "Behind Quantum Electronics," 150, 216, 219.

11. Kevles, "Cold War," 264.

12. David A. Hounshell, "Rethinking the Cold War; Rethinking Science and Technology in the Cold War; Rethinking the Social Study of Science and Technology," *Social Studies of Science* 31 (2001): 289–297, on pp. 290–291.

13. The text of the bill is in Richard G Hewlett and Oscar E. Anderson, Jr., eds., *A History of the United States Atomic Energy Commission*, vol. 1: *The New World, 1939–1946* (University Park: Pennsylvania State University, 1962), 425–427. See Alice Kimball Smith, *A Peril and a Hope: The Scientists' Movement in America, 1945–1947* (Cambridge, MA: MIT Press, 1970), chap. 3.

14. William S. White, "Scientist Group Hits Atomic Bill," *New York Times*, October 19, 1945, 2; letter of Bush to Don K. Price, 30 December 1953. Vannevar Bush papers, box 94, folder 2147. Manuscripts Division, U.S. Library of Congress, Washington, DC.

15. Text published in Robert C. Williams and Philip L. Cantelon, eds., *The American Atom: A Documentary History of Nuclear Policies from the Discovery of Fission to the Present, 1939–1984* (Philadelphia: University of Pennsylvania Press, 1984), 79–92.

16. See Richard T. Sylves, *The Nuclear Oracles: A Political History of the General Advisory Committee of the Atomic Energy Commission, 1947–1977* (Ames: Iowa State University Press, 1987).

17. Kevles, *The Physicists*, 374; Seidel, "Accelerating Science"; Robert W. Seidel, "The National Laboratories of the Atomic Energy Commission in the Early Cold War," *Historical Studies in the Physical and Biological Sciences* 32, no. 1 (2001): 145–162; and Robert Crease, *Making Physics: A Biography of Brookhaven National Laboratory, 1946–1972* (Chicago: University of Chicago Press, 1999).

18. Armin Hermann, John Krige, Ulrike Mersits, and Dominique Pestre, *History of CERN*, vol. 1: *Launching the European Organization for Nuclear Research* (Amsterdam: North Holland, 1987); John Krige, "I. I. Rabi and CERN," *Physics*

in Perspective 7 (2005): 150–164; and John Rigden, *Rabi: Scientist and Citizen* (New York: Basic Books, 1987), 232–241.

19. John Krige, *American Hegemony and the Postwar Reconstruction of Science in Europe* (Cambridge, MA: MIT Press, 2006), 69–71, 257–258; and "Atoms for Peace, Scientific Internationalism and Scientific Intelligence," *Osiris* 21 (2006): 161–181.

20. The Atoms for Peace program is further discussed in Richard G. Hewlett and Jack M. Holl, *Atoms for Peace and War, 1953–1961: Eisenhower and the Atomic Energy Commission* (Berkeley: University of California Press, 1989); John Krige, *American Hegemony;* and Krige, "Atoms for Peace."

21. "President Eisenhower's 'Atoms for Peace' Speech, December 8, 1953," www.atomicarchive.com/Docs/Deterrence/Atomsforpeace.shtml.

22. Krige, "Atoms for Peace."

23. Quoted by Krige, ibid., 180.

24. D. C. Cassidy, *J. Robert Oppenheimer and the American Century* (New York: Pi Press, 2005; repr., Baltimore: Johns Hopkins University Press, 2009), appendix 2; David Kaiser, "Cold War Requisitions, Scientific Manpower, and the Production of American Physicists after World War II," *Historical Studies in the Physical and Biological Sciences* 33, no. 1 (2002): 131–159.

25. Cassidy, *J. Robert Oppenheimer,* appendix 1. The other eight were Harvard, Connecticut, Illinois, Yale, Cornell, Chicago, Caltech, and Michigan.

26. Quoted in Margaret W. Rossiter, *Women Scientists in America: Before Affirmative Action, 1940–1972* (Baltimore: Johns Hopkins University Press, 1995), 53.

27. Further discussed by Rossiter, ibid., chap. 3.

28. Ibid., tables on pp. 81–82,86–87.

29. Based on data provided by Rossiter, ibid., tables on pp. 109, 188–189, 259.

6. THE NEW PHYSICS

1. Spencer R. Weart, "The Solid Community," in *Out of the Crystal Maze: Chapters from the History of Solid-State Physics,* ed. Lillian Hoddeson, Ernest Braun, Jürgen Teichmann, and Spencer Weart (New York: Oxford University Press, 1992), 617–669, on p. 641.

2. Silvan S. Schweber, *QED and the Men Who Made It: Dyson, Feynman, Schwinger, and Tomonaga* (Princeton, NJ: Princeton University Press, 1994), 153.

3. Weart, "Solid Community."

4. Vivian A. Johnson, *Karl Lark-Horovitz: Pioneer of Solid State Physics* (New York: Pergamon Press, 1969), 36. Purdue's work is discussed, in addition, by Paul W. Hendriksen, "Solid State Physics Research at Purdue," *Osiris* 3 (1987): 237–260; and Ralph Bray, "The Origin of Semiconductor Research at Purdue," *Purdue Physics,* Newsletter of the Department of Physics, College of Science, Purdue University (1989), at www.physics.purdue.edu/about_us/history/semi_conductor _research.shtml.

5. Quoted by Weart, "Solid Community," 645.

6. For more on the transistor, see Lillian Hoddeson, "The Discovery of the Point-Contact Transistor," *Historical Studies in the Physical Sciences* 12, no. 1 (1981):

41–76; and Lillian Hoddeson and Michael Riordan, *Crystal Fire: The Invention of the Transistor and the Birth of the Information Age* (New York: W. W. Norton, 1997). For a general introduction, see *Transistorized!*, directed by Gary Glassman (1999; Alexandria, VA: PBS Home Video, 1999), DVD; and the related website www.pbs.org/transistor.

7. See Hoddeson and Riordan, *Crystal Fire.*

8. There is a large literature on the history of the maser and laser, and their military context. It includes Joan Bromberg, *The Laser in America, 1950–1970* (Cambridge, MA: MIT Press, 1991); J. Hecht, *Beam: the Race to Make the Laser* (New York: Oxford University Press, 2005); Donald F. Nelson, Robert J. Collins, and Wolfgang Kaiser, "Bell Labs and the Ruby Laser," *Physics Today* 63 (January 2010): 40–45; Paul Forman, "Inventing the Maser in Postwar America," *Osiris* 7 (1992): 105–134; and Paul Forman, "Into Quantum Electronics: The Maser as 'Gadget' of Cold-War America," in *National Military Establishments and the Advancement of Science and Technology: Studies in 20th Century History*, ed. Paul Forman and J. Sánchez-Ron (Dordrecht: Kluwer Academic, 1996), 261–326. For general readers, see also "Bright Idea: The First Lasers," at www.aip.org/history/exhibits/laser.

9. Charles H. Townes," Science and Technology," in Henry H. Stroke, ed., *The* Physical Review: *The First Hundred Years, A Selection of Seminal Papers and Commentaries* (Woodbury, NY: American Institute of Physics Press, 1995); Robert V. Pound, "Edward Mills Purcell," in *Biographical Memoirs of the National Academy of Sciences*, vol. 78 (Washington, DC: National Academies Press, 2000), also at www.nap.edu/readingroom.php?book=biomems&page=epurcell.html; Solomon Gartenhaus, "Albert W. Overhauser," Purdue University, College of Science, Department of Physics, www.physics.purdue.edu/about_us/history/Albert_W_Overhauser.shtml.

10. Silvan S. Schweber, "Shelter Island, Pocono, and Oldstone: The Emergence of American Quantum Electrodynamics after World War II," *Osiris* 2 (1986): 265–302, on p. 266. See also Schweber, *QED;* and Abraham Pais, *Inward Bound: Of Matter and Forces in the Physical World* (New York: Oxford University Press, 1988), chap. 18.

11. For more details, see the works in note 10, as well as David Kaiser, *Drawing Theories Apart: The Dispersion of Feynman Diagrams in Postwar Physics* (Chicago: University of Chicago Press, 2005).

12. Laurie M. Brown, Max Dresden, and Lillian Hoddeson, eds., *Pions to Quarks: Particle Physics in the 1950s: Based on a Fermilab Symposium* (New York: Cambridge University Press, 1989).

13. Daniel J. Kevles, *The Physicists: The History of a Scientific Community in Modern America* (New York: Vintage Books, 1971; repr. Cambridge, MA: Harvard University Press, 1995), 366.

14. Don K. Price, *The Scientific Estate* (Cambridge, MA: Belknap Press of Harvard University Press, 1965), 3.

15. Mary M. Simpson, "The Scientist in Politics: On Top or on Tap?," *Bulletin of the Atomic Scientists* 16 (1960): 28–29. A survey of the literature on these issues

through 1992 is offered by James H. Capshew and Karen A. Rader, "Big Science: Price to the Present," *Osiris* 7 (1992): 3–25.

16. Reported in "War Ban Is Urgent, 515 Scientists Say," *New York Times*, October 31, 1946, 6.

17. Oppenheimer to Lawrence, 30 August 1945, in *Robert Oppenheimer: Letters and Recollections*, ed. Alice Kimball Smith and Charles Weiner (Cambridge, MA: Harvard University Press, 1980), 301; Barton Bernstein, "Four Physicists and the Bomb: The Early Years 1945–1950," *Historical Studies in the Physical Sciences* 18, no. 2 (1988): 231–263, p. 242. For more on the early scientists' movement, see Alice Kimball Smith, *A Peril and a Hope: The Scientists' Movement in America, 1945–1947* (Cambridge, MA: MIT Press, 1970), chap. 11; and Robert Gilpin, *American Scientists and Nuclear Weapons* (Princeton, NJ: Princeton University Press, 1962).

18. "Truman Statement on Atom," *New York Times*, September 24, 1949, 1.

19. See Edward Teller and Judith Schoolery, *Memoirs: A Twentieth-Century Journey in Science and Politics* (New York: Basic Books, 2002).

20. Teller, memo of October 13, 1949, quoted by James W. Kunetka, *Oppenheimer: The Years of Risk* (Englewood Cliffs, NJ: Prentice-Hall, 1982), 151; Einstein, from the original German, in Albert Einstein, *Über den Frieden*, ed. Otto Nathan and Heinz Norton (Zurich: Book Club ex libris, 1976), 519, translated as "This technological and psychological orientation in military policy" in Albert Einstein, *Einstein on Peace*, ed. Otto Nathan and Heinz Norden (New York: Schocken Books, 1960), 521.

21. Reported by Lilienthal, journal entry for October 29, 1949, in David E. Lilienthal, *The Journals of David E. Lilienthal*, vol. 2: *The Atomic Energy Years, 1945–1950* (New York: Harper and Row, 1964), 581–582. The General Advisory Committee deliberations and the decision to build the H-bomb are discussed, by, among others, Richard Rhodes, *Dark Sun: The Making of the Hydrogen Bomb* (New York: Simon and Schuster, 1995); Peter Galison and Barton Bernstein, "In Any Light: Scientists and the Decision to Build the Superbomb, 1952–1954," *Historical Studies in the Physical Sciences* 19, no. 2 (1989): 267–347; and McGeorge Bundy, *Danger and Survival: Choices about the Bomb in the First Fifty Years* (New York: Random House, 1988).

22. Oppenheimer to Lilienthal, report on 17th meeting of General Advisory Committee, October 29–30, 1949, dated October 30, 1949, original at the Department of Energy, Las Vegas, NV, repository, 0104151, published with appendices in Robert C. Williams and Philip L. Cantelon, eds., *The American Atom: A Documentary History of Nuclear Policies from the Discovery of Fission to the Present, 1939–1984* (Philadelphia: University of Pennsylvania Press, 1984), 120–127.

23. Ibid.

24. Edward Teller, "Back to the Laboratories," *Bulletin of the Atomic Scientists* 6 (1950): 71–72.

25. "Let Us Pledge Not to Use the H-Bomb First!" *Bulletin of the Atomic Scientists* 6 (1950): 71–72, petition distributed to the American Physical Society meeting in New York, January 1950, with twelve signatures, reported in *New York Times*, February 5, 1950, 1, 3.

26. Silvan S. Schweber, *In the Shadow of the Bomb: Oppenheimer, Bethe and the Moral Responsibility of the Scientist* (Princeton, NJ: Princeton University Press, 2000), 163–164. The phrase appeared in President John F. Kennedy's inaugural address of 1961.

27. Karl Hufbauer, "J. Robert Oppenheimer's Path to Black Holes," in *Reappraising Oppenheimer: Centennial Studies and Reflections*, ed. Cathryn Carson and David A. Hollinger (Berkeley: University of California Press, 2005), 31–47; J. A. E. F van Dongen, "Black Hole Interpretations (1916–1996)" (master's thesis, Universiteit van Amsterdam, 1998); and D. C. Cassidy, *J. Robert Oppenheimer and the American Century* (New York: Pi Press, 2005; repr., Baltimore: Johns Hopkins University Press, 2009), 173–178.

28. The development of the H-bomb is discussed by Richard Rhodes, *Dark Sun: The Making of the Hydrogen Bomb* (New York: Simon and Schuster, 1995), esp. 466–467. Oppenheimer's remark may be found in U.S. Atomic Energy Commission (AEC), *In the Matter of J. Robert Oppenheimer: Transcript of Hearing before Personnel Security Board and Texts of Principal Documents and Letters* (Cambridge, MA: MIT Press, 1954, repr., 1970), 251.

29. "Announced Nuclear Tests, 1945–52," in Williams and Cantelon, *American Atom*, 181.

30. "The Hidden Struggle for the H-Bomb: The Story of Dr. Oppenheimer's Persistent Campaign to Reverse U.S. Military Strategy," *Fortune*, May 1953, 109–110, 230. The article was published anonymously.

31. Nichols, letter to Oppenheimer, December 23, 1953, published in AEC, *In the Matter*, 3–7.

32. These include Priscilla McMillan, *The Ruin of J. Robert Oppenheimer and the Birth of the Modern Arms Race* (New York: Viking, 2005); Philip M. Stern and Harold P. Green, *The Oppenheimer Case: Security on Trial* (New York: Harper and Row, 1969); Barton J. Bernstein, "In the Matter of J. Robert Oppenheimer," *Historical Studies in the Physical Sciences* 12, no. 2 (1982): 195–252; John Major, *The Oppenheimer Hearing* (New York: Stein and Day, 1971); Rachel L. Holloway, *In the Matter of J. Robert Oppenheimer: Politics, Rhetoric, and Self-Defense* (Westport, CT: Praeger, 1993); and Heinar Kipphardt, *In the Matter of J. Robert Oppenheimer: A Play Freely Adapted on the Basis of the Documents*, trans. Ruth Speirs (New York: Hill and Wang, 1964).

33. Rabi, quoted in AEC, *In the Matter*, 468.

34. Personnel Security Board, majority report, in AEC, *In the Matter*, 999–1019, on pp. 1016–1018.

35. Mrs. Cronk to DuBridge, 30 June 1954, and C. Walsh to DuBridge, 3 July 1954. Lee A. DuBridge papers, box 33, folders 2 and 5. California Institute of Technology Archives, Pasadena, CA.

36. Richard G. Hewlett and Jack M. Holl, *Atoms for Peace and War, 1953–1961: Eisenhower and the Atomic Energy Commission* (Berkeley: University of California Press, 1989), 111–112.

37. Paul Forman, "Behind Quantum Electronics: National Security as Basis for Physical Research in the United States, 1940–1960," *Historical Studies in the Physical Sciences* 18, no. 1 (1987): 149–229, on p. 29.

7. SPUTNIK

1. Quotations from the minutes by John S. Rigden, "Eisenhower, Scientists, and *Sputnik*," *Physics Today* 60, no. 6 (June 2007): 47–52, on p. 50.

2. Bruce L. R. Smith, *The Advisors: Scientists in the Policy Process* (Washington, DC: Brookings Institution, 1992), 165. See also Zuoyue Wang, *In Sputnik's Shadow: The President's Science Advisory Committee and Cold War America* (New Brunswick, NJ: Rutgers University Press, 2008).

3. "Text of the Address by President Eisenhower on Science in National Security," *New York Times*, November 8, 1957, 10.

4. See also Patrick J. Mulvey and Starr Nicholson, "Enrollments and Degrees Report, 2006," *AIP Report*, American Institute of Physics, Statistical Research Center, no. R-151.43 (September 2008), 16, fig. 13; David Kaiser, "Cold War Requisitions, Scientific Manpower, and the Production of American Physicists after World War II," *Historical Studies in the Physical and Biological Sciences* 33, no. 1 (2002): 131–159, esp. p. 135, fig. 2.

5. See Daniel J. Kevles, *The Physicists: The History of a Scientific Community in Modern America* (New York: Vintage Books, 1971; repr. Cambridge, MA: Harvard University Press, 1995), chap. 23.

6. National Science Foundation (NSF), "Federal Funds for R&D," NSF 01–334 statistical tables, "Table A. Federal Obligations for Research and Development, by Character of Work, R&D Plant, and Major Agency: Fiscal Years 1951–2002," at www.nsf.gov/statistics/fedfunds.

7. Ibid.; also U.S. Office of Budget and Management, *Historical Tables: Budget of the U.S. Government, Fiscal Year 2011* (Washington, DC: Government Printing Office, 2010), "Table 1-1: Summary of Receipts, Outlays, and Surpluses or Deficits: 1789–2015," www.gpoaccess.gov/usbudget/fy11/pdf/hist.pdf. The inflation adjustment is based on the consumer price index calculator provided by the U.S. Bureau of Labor Statistics at www.bls.gov/data/inflation_calculator.htm.

8. The *Physical Review* Online Archive (PROLA) is found at http://prola.aps .org. See Paul Hartman, *A Memoir on the* Physical Review: *A History of the First Hundred Years* (Woodbury, NY: American Institute of Physics, 1994), 182–183.

9. Hartman, *A Memoir*, 188.

10. Lillian Hoddeson, Adrienne W. Kolb, and Catherine Westfall, *Fermilab: Physics, the Frontier, and Megascience* (Chicago: University of Chicago Press, 2008), chap. 1.

11. Kevles, *The Physicists*, chap. 23.

12. Hoddeson et al., *Fermilab*, chap. 2; Peter J. Westwick, *The National Labs: Science in an American System, 1947–1974* (Cambridge, MA: Harvard University Press, 2003), 165–172.

13. Westwick, "National Labs," 182–183; Daniel S. Greenberg, *The Politics of Pure Science*, 2nd ed. (Chicago: University of Chicago Press, 1999), 232–239; W. K. H. Panofsky, "SLAC and Big Science: Stanford University," in *Big Science: The Growth of Large-Scale Research*, ed. Peter Galison and Bruce W. Hevly (Stanford, CA: Stanford University Press, 1992), 129–146; Peter Galison, Bruce W.

Hevly, and Rebecca Lowen, "Controlling the Monster: Stanford and the Growth of Physics Research, 1935–1962," in Galison and Hevly, *Big Science*, 46–77; Zuoyue Wang, "The Politics of Big Science in the Cold War: PSAC and the Funding of SLAC," *Historical Studies in the Physical Sciences* 25, no. 2 (1995): 329–356.

14. Quoted by Hoddeson et al., *Fermilab*, 37. The panel was chaired by physicist Norman Ramsey.

15. Alvin M. Weinberg, "The Choices of Big Science: 1. Criteria for Scientific Choice" [1963], in *Reflections on Big Science* (Cambridge, MA: MIT Press, 1967), 65–84, passage from pp. 78–80.

16. Lillian Hoddeson, Helmut Schubert, Steve J. Heims, and Gordon Baym, "Collective Phenomena," in *Out of the Crystal Maze: Chapters from the History of Solid-State Physics*, ed. Lillian Hoddeson, Ernest Braun, Jürgen Teichmann, and Spencer Weart (New York: Oxford University Press, 1992), 489–616, esp. pp. 541–564; and Gloria Lubkin, "Nobel Prize Shared by Esaki, Giaever and Josephson," *Physics Today* 26 (December 1973): 73–75, citing an interview with Anderson.

17. The history of Fermilab is nicely recounted by Hoddeson et al., *Fermilab*.

18. Quoted by Gloria B. Lubkin, "Nobel Prizes: to Glashow, Salam and Weinberg for Physics," *Physics Today* 32 (December 1979): 17–19, on p. 19. On the discovery of the top quark, see Ray Ladbury, "Where Do You Go When You've Made It to the Top?" *Physics Today* 48 (May 1995): 17–19. For further details on this work, see Abraham Pais, *Inward Bound: Of Matter and Forces in the Physical World* (New York: Oxford University Press, 1988); Lillian Hoddeson, Laurie Brown, Michael Riordan, Max Dresden, eds., *The Rise of the Standard Model: Particle Physics in the 1960s and 1970s* (New York: Cambridge University Press, 1997). An excellent though brief nontechnical summary is offered by Graham Farmelo, *The Strangest Man: The Hidden Life of Paul Dirac, Mystic of the Atom* (New York: Basic Books, 2009), 381–382, 398–400.

19. See Paul Halpern, *Collider: The Search for the World's Smallest Particles* (New York: John Wiley and Sons, 2009).

20. The politics of science in general and of physics in the 1960s and early 1970s are further discussed by Smith, *The Advisors*; Greenberg, *Politics of Pure Science*; Kevles, *The Physicists*, chaps. 24–25; and Kelly Moore, *Disrupting Science: Social Movements, American Scientists, and the Politics of the Military, 1945–1975* (Princeton, NJ: Princeton University Press, 2008).

21. Donna Miles, "U.S. Declassifies Nuclear Stockpile Details to Promote Transparency," Department of Defense, *Armed Forces Press Service*, May 3, 2010, www.defense.gov/news/newsarticle.aspx?id=59004; Hans M. Kristensen, "United States Discloses Size of Nuclear Weapons Stockpile," Federation of American Scientists, *Strategic Security Blog*, May 3, 2010, www.fas.org/blog/ssp/2010/05/stockpilenumber.php.

22. Smith, *The Advisors*, 167.

23. Raised in Don K. Price, *The Scientific Estate* (Cambridge, MA: Belknap Press of Harvard University Press, 1965), 3; Smith, *The Advisors*, 168.

24. Quoted in his report on a Congressional hearing on the NSF budget: D. S. Greenberg, "NSF Budget: Cuts by House Group Leave Little Leeway for Growth

in Support of Research Projects," *Science* 148, no. 3672 (1965): 928–930; see also data from Kevles, *The Physicists*, 396–397, and NSF, "Federal Funds for R&D," NSF 01–334, Table A.

25. See also Greenberg, *Politics of Pure Science*, 210.

26. NSF, "Federal Funds for R&D," NSF 01–334, table A; U.S. Office of Budget and Management, *Historical Tables*, table 1-1.

27. Kaiser, "Cold War Requisitions," 151.

28. On the Society for the Social Responsibility of Science, see Moore, *Disrupting Science*, chap. 3.

29. "Founding Document: 1968 MIT Faculty Statement," on the Union of Concerned Scientists website, www.ucsusa.org/about/founding-document-1968.html.

30. Moore, *Disrupting Science*, chap. 5.

31. Gloria B. Lubkin, conversation with the author, May 25, 2010; Barry M. Casper, "Physicists and Public Policy: The 'Forum' and the APS," *Physics Today* 27 (May 1974): 31–37; and Gloria B. Lubkin, "Discrimination against Women in Physics," *Physics Today* 25 (July 1972): 61–62.

32. NSF, "Federal Funds for R&D," NSF 01–334, table A.

33. Kevles, *The Physicists*, 415.

34. Paul Forman, "Alfred Landé and the Anomalous Zeeman Effect, 1919–1921," *Historical Studies in the Physical Sciences* 2 (1970): 153–261, on p. 157. Protester quoted by Moore, *Disrupting Science*, 167. The origins of this approach are further discussed by David C. Cassidy, "Paul Forman and the Environment and Practice of Quantum History," in Paul Forman, *Weimar Culture and Quantum Mechanics: Selected Papers by Paul Forman and Contemporary Perspectives on the Forman Thesis*, ed. Cathryn Carson et al. (London: Imperial College Press, forthcoming).

35. Quoted by Robert Reinhold, "Scientists Urged to Stay Relevant," *New York Times*, November 2, 1969, 31; and Kevles, *The Physicists*, 412.

36. This point is further discussed by Smith, *The Advisors*, 172–176.

37. Ibid., 173.

38. For further discussion, see Mel Horwitch, *Clipped Wings: The American SST Conflict* (Cambridge, MA: MIT Press, 1982).

8. REVISING THE PARTNERSHIP

1. U.S. Office of Budget and Management, *Historical Tables: Budget of the U.S. Government, Fiscal Year 2011* (Washington, DC: US Government Printing Office, 2010), table 4.1, "Outlays by Agency: 1962–2015," www.gpoaccess.gov/usbudget/fy11/pdf/hist.pdf; Patrick J. Mulvey and Starr Nicholson, "Enrollments and Degrees Report, 2006," American Institute of Physics, Statistical Research Center, *AIP Report*, R-151.43 (September 2008), 16, fig. 13.

2. Bruce L. R. Smith, *The Advisors: Scientists in the Policy Process* (Washington, DC: Brookings Institution, 1992), 178.

3. In 1962, Atomic Energy Commission funding for R&D amounted to about 11.9 percent of the federal outlays for R&D (Table 2 in the Appendix). The

inflation adjustment is based on the consumer price index calculator provided by the U.S. Bureau of Labor Statistics at www.bls.gov/data/inflation_calculator .htm.

4. National Science Foundation (NSF), Division of Science Resources Statistics, *Science and Engineering Degrees: 1966–2000*, NSF 02-327, ed. Susan T. Hill (Arlington, VA: 2000), table 38, www.nsf.gov/statistics/nsf02327.

5. R. Joseph Anderson and Orville R. Butler, with M. Juris, "History of Physicists in Industry: Final Report" (College Park, MD: American Institute of Physics, October 2008), 4, www.aip.org/history/pubs/HOPI_Final_report.pdf.

6. Ibid., 8–9.

7. "Alcatel, Telcos, Chip Makers Launch Green Research," *New York Times*, January 11, 2010.

8. Thomas Friedman, *The Earth Is Flat: A Brief History of the Twenty-First Century* (New York: Farrar, Straus, and Giroux, 2006).

9. Anderson, Butler, Juris, note 5; and Kate Kirby, Roman Czujko, Patrick Mulvey, "The Physics Job Market: From Bear to Bull in a Decade," *Physics Today* 54 (April 2001): 36–41, on p. 37.

10. NSF, *Science and Engineering Degrees: 1966–2000*; Patrick J. Mulvey, "Number of PhDs Granted in Selected Subfields, 2000," *Physics Trends*, Fall 2002, www .aip.org/statistics/trends/reports/fall2002c.pdf; and American Physical Society, "Official 2009 Unit Membership Statistics," attachment 1, showing 2005–2009, at www.aps.org/units/dfd/governance/minutes/upload/spring09.pdf (accessed January 2011).

11. For the former, see Spencer R. Weart, "Global Warming, Cold War, and the Evolution of Research Plans," *Historical Studies in the Physical and Biological Sciences* 27, no. 2 (1997): 319–356; and Spencer R. Weart, *The Discovery of Global Warming* (Cambridge, MA: Harvard University Press, 2003). For the latter, see Lawrence Badash, *A Nuclear Winter's Tale: Science and Politics in the 1980s* (Cambridge, MA: MIT Press, 2009).

12. For instance, "AEC Plans Computer Center for Fusion," *Physics Today* 27 (October 1974): 19–20. See Robert W. Seidel, "From Mars to Minerva: The Origins of Scientific Computing in the AEC Labs," *Physics Today* 49, no. 10 (October 1996): 33–39.

13. Herbert H. Goldstine, *The Computer: From Pascal to von Neumann* (Princeton, NJ: Princeton University Press, 1972), part 3, chap. 5.

14. Seidel, "From Mars to Minerva"; and Alfred E. Brenner, "The Computing Revolution and the Physics Community," *Physics Today* 49, no. 10 (October 1996): 24–30.

15. Peter Galison, *Image and Logic: A Material Culture of Microphysics* (Chicago: University of Chicago Press, 1997), chap. 5.

16. Galison, *Image and Logic*, 313–320; Robert W. Seidel, "From Factory to Farm: Dissemination of Computing in High-Energy Physics," *Historical Studies in the Natural Sciences* 38, no. 4 (2008): 479–507.

17. Charles J. Murray, *The Supermen: The Story of Seymour Cray and the Technical Wizards behind the Super Computer* (New York: John Wiley and Sons, 1997). As of

2011, the world's most powerful computer, the Chinese-built Tianhe-1A, could perform 2,500 billion Mflops, or 2.5 petaflops.

18. This is argued by Brenner, "Computing Revolution," 28. For one use of computers as a research tool, see Mitchell J. Feigenbaum, "Computer-Generated Physics," in *Twentieth Century Physics*, 3 vols., ed. Abraham Pais, A. B. Pippard, and Laurie M. Brown (Bristol and Philadelphia: IOP Press; New York: AIP Press, 1995), 3: 1823–1853.

19. Seidel, "From Factory to Farm."

20. Glenn E. Bugos, *Atmosphere of Freedom: Sixty Years at the NASA Ames Research Center* (Washington, DC: NASA History Office, 2000), 109.

21. R. P. Turco, O. B. Toon, T. P. Ackerman, J. B. Pollack, and Carl Sagan, "Nuclear Winter: Global Consequences of Multiple Nuclear Explosions," *Science* 222, no. 4630 (1983): 1283–1289. For further discussion of the TTAPS work, see Badash, *Nuclear Winter's Tale.*

22. Paul R. Ehrlich, J. Harte, M. A. Harwell, P. H. Raven, C. Sagan, G. M. Woodwell, et al., "Long-Term Biological Consequences of Nuclear War," *Science* 222, no. 4630 (1983): 1293–1300.

23. See Badash, *Nuclear Winter's Tale*, for further discussion.

24. Francis Fitzgerald, *Way Out There in the Blue: Reagan, Star Wars, and the End of the Cold War* (New York: Simon and Schuster, 2000); William J. Broad, *Teller's War: The Top-Secret Story behind the Star Wars Deception* (New York: Touchstone Books, 1993).

25. Quoted from the abstract by Gloria B. Lubkin, "APS Releases Report on Directed-Energy Weapons," *Physics Today* 40 (May 1987), S1–S2. Full report found in N. Bloembergen, C. K. N. Patel, P. Avizonis, R. G. Clem, A. Hertzberg, T. H. Johnson, et al., "Report to the American Physical Society of the Study Group on Science and Technology of Directed Energy Weapons," *Reviews of Modern Physics* 59, no. 3 (1987), S1–S202.

26. R. P. Turco, O. B. Toon, T. P. Ackerman, J. B. Pollack, and C. Sagan, "Climate and Smoke: An Appraisal of Nuclear Winter," *Science* 247, no. 247 (1990): 166–176, on p. 167–168.

27. O. B. Toon, R. P. Turco, A. Robock, C. Bardeen, L. Oman, and G. L. Stenchikov, "Atmospheric Effects and Societal Consequences of Regional Scale Nuclear Conflicts and Acts of Individual Terrorism," *Atmospheric Chemistry and Physics Discussions* 6 (2006): 11745–11816; A. Robock, L. Oman, G. L. Stenchikov, O. B. Toon, C. Bardeen, and R. P. Turco, "Climactic Consequences of Regional Nuclear Conflicts," *Atmospheric Chemistry and Physics Discussions* 7 (2007): 2003–2012; Alan Robock and Owen Brian Toon, "South Asia Threat? Local Nuclear War = Global Suffering," *Scientific American* 302 (2010): 74–81.

28. Lillian Hoddeson, Adrienne W. Kolb, and Catherine Westfall, *Fermilab: Physics, the Frontier, and Megascience* (Chicago: University of Chicago Press, 2008), chap. 10.

29. Ibid., chap. 12.

30. Sheldon L. Glashow and Leon M. Lederman, "The SSC: A Machine for the Nineties," *Physics Today* 38, no. 3 (1985): 28–37, on p. 34.

31. Authoritative histories of the rise and fall of the SSC include, in chronological order, Daniel J. Kevles, "Preface, 1995: The Death of the Superconducting Super Collider in the Life of American Physics," in Daniel J. Kevles, *The Physicists: The History of a Scientific Community in Modern America* (New York: Vintage Books, 1971; repr. with new preface, Cambridge, MA: Harvard University Press, 1995), ix–xlii; Lillian Hoddeson and Adrienne W. Kolb, "The Superconducting Super Collider's Frontier Outpost, 1983–1988," *Minerva* 38 (2000): 271–310; Michael Riordan, "The Demise of the Superconducting Super Collider," *Physics in Perspective* 2 (2000): 411–425; Michael Riordan, "A Tale of Two Cultures: Building the Superconducting Super Collider, 1988–1993," *Historical Studies in the Physical and Biological Sciences* 32, no. 1 (2001): 125–144; and Hoddeson et al., *Fermilab*, chap. 13.

32. U.S. Office of Budget and Management, *Historical Tables*, "Table 1-1: Summary of Receipts, Outlays, and Surpluses or Deficits: 1789–2015."

33. Quoted by Kevles, "Preface," xxix.

34. Ibid., xxv.

35. Ibid., xli.

36. John Horgan, *The End of Science: Facing the Limits of Knowledge in the Twilight of the Scientific Age* (New York: Broadway Books, 1997).

37. For instance, in that decade, Peter Galison, *Image and Logic: A Material Culture of Microphysics* (Chicago: University of Chicago Press, 1997); Andrew Pickering, *Constructing Quarks: A Sociological History of Particle Physics* (Chicago: University of Chicago Press, 1999).

38. Paul Forman, "Independence, Not Transcendence, for the Historian of Science," *Isis* 82 (1991): 71–86; and Paul Forman, "Transcendence, or the Flight from Responsibility: Modern Science in Postmodern Perspective" (paper presented at the conference "Scienza e Potere," Florence, Italy, 1994). I thank Paul Forman for providing a copy of his manuscript.

39. Ibid., 2.

40. Ibid., 1. See also, more recently, Paul Forman, "(Re)cognizing Postmodernity: Helps for Historians—of Science Especially," *Berichte zur Wissenschaftsgeschichte* 33 (2010): 157–175. These and other works by Forman may be included in his forthcoming collection: Paul Forman, *Against Transcendence: Essays on the Physical Significance of Modern Culture* (New York: Cambridge University Press, scheduled for release May 30, 2011).

41. John A. Wheeler, "Our Universe: The Known and the Unknown," *American Scientist* 1 (1968): 1–20, on 8–9, address to American Association for the Advancement of Science, December 1967.

42. See Alan H. Guth, *The Inflationary Universe: The Quest for a New Theory of Cosmic Origins* (New York: Perseus Books, 1997).

43. For more on cosmology and related developments, see, among the many works on these subjects, Stephen Hawking, *A Brief History of Time* (New York: Bantam Books, 1988, repr. 1996); Mario Livio, *The Accelerating Universe: Infinite Expansion, the Cosmological Constant, and the Beauty of the Cosmos* (New York: Wiley, 2000); and Helge Kragh, *Conceptions of the Universe: From Myths to the Accelerating Universe* (New York: Oxford University Press, 2007).

44. Dennis Overbye, "With a Mighty Smash, Europe Seizes the Lead in Big Physics," *New York Times*, December 10, 2009, A1, A4.

45. Jeffrey Mervis, "Handful of U.S. Schools Claim Larger Share of Output," *Science* 330, no. 6007 (2010): 1032.

46. Purdue University, *Purdue University Data Digest*, www.purdue.edu/Data Digest.

47. Data based on Patrick J. Mulvey and Starr Nicholson, "Physics Undergraduate Enrollments and Degrees: Results from the 2007 Survey of Enrollments and Degrees," American Institute of Physics, Statistical Research Center, *Focus On*, R-151.44 (January 2010), 6, fig. 4; 10, fig. 7; 16, table 6, www.aip.org/statistics/trends/reports/EDphysund07.pdf; National Science Foundation (NSF), *Women, Minorities, and Persons with Disabilities in Science and Engineering, Doctoral degrees*, F-1, "S&E Doctoral Degrees Awarded, by Field: 2000–2008," and F-2, "S&E Doctoral Degrees Awarded to Women, by Field: 2000–2008," data updated October 2010, www.nsf.gov/statistics/wmpd/degrees.cfm#doctoral.

48. American Physical Society, "Official 2010 Unit Member Statistics," www.aps.org/membership/units/upload/YearlyUnit10.pdf (accessed January 2011), showing 2006–2010.

Acknowledgments

In addition to the many authors whose works have helped enlighten this account, I am very grateful to Spencer R. Weart, who provided invaluable advice and insight throughout the course of this project. I am also very grateful to Martin Blume, Gloria B. Lubkin, Peter Pesic, Elizabeth G. Knoll, and anonymous referees for their many helpful insights, comments, and suggestions, and to the Department of Physics at Purdue University. I would like to thank Spencer R. Weart and Margaret C. Jacob, the series editors, and Elizabeth G. Knoll and the staff of Harvard University Press for including this book in the series New Histories of Science, Technology, and Medicine, and for their fine work in bringing it into print. I wish to thank Hofstra University for release time from teaching and for a special research leave that made the completion of this work possible.

Index